"A STUDY ON TRIBAL FARMERS OF SURGUJA DIVISION WITH REFERENCE TO ADOPTION OF SUGARCANE PRODUCTION TECHNOLOGY"

M.Sc. (Ag.) THESIS

BY

RAHUL KUMAR TIWARI

DEPARTMENT OF AGRICULTURAL EXTENSION

COLLEGE OF AGRICULTURE
INDIRA GANDHI KRISHI VISHWAVIDYALAYA

RAIPUR (C.G.)

2014

"A STUDY ON TRIBAL FARMERS OF SURGUJA DIVISION WITH REFERENCE TO ADOPTION OF SUGARCANE PRODUCTION TECHNOLOGY"

THESIS

SUBMITTED TO THE

Indira Gandhi Krishi Vishwavidyalaya, Raipur

By

RAHUL KUMAR TIWARI

IN PARTIAL FULFILMENT OF THE REQUIREMENTS FOR THE DEGREE
OF

Master of Science

In

Agriculture

(AGRICULTURAL EXTENSION)

Roll No. 15538   ID No. 110107040

JULY, 2014

CERTIFICATE -I

This is to certify that the thesis entitled "A STUDY ON TRIBAL FARMERS OF SURGUJA DIVISION WITH REFERENCE TO ADOPTION OF SUGARACANE PRODUCTION TECHNOLOGY" submitted in partial fulfillment of the requirements for the degree of "MASTER OF SCIENCE IN AGRICULTURE" of the Indira Gandhi Krishi Vishwavidyalaya, Raipur is a record of the bonafide research work carried out by **Shri RAHUL KUMAR TIWARI** under the guidance and supervision. The subject of the thesis has been provided by Student's Advisory committee and the Director of Instructions.

No part of the thesis has been submitted for any other degree or diploma (certificate awarded etc.) or has been published /published part has been fully acknowledged. All the assistance and help received during the course of investigation has been duly acknowledged by him.

Date

Chairman
Advisory Committee

THESIS APPROVED BY THE STUDENT'S ADVISORY COMMITTEE

Chairman Dr. P.K.Jaiswal

Member Dr.M.A. Khan

Member Dr.A.K.Sinha

Member Smt.N.Chaoksey

CERTIFICATE -II

This is to certify that the thesis entitled "A STUDY ON TRIBAL FARMERS OF SURGUJA DIVISION WITH REFERENCE TO ADOPTION OF SUGARACANE PRODUCTION TECHNOLOGY" submitted by Shri RAHUL KUMAR TIWARI to the Indira Gandhi Krishi Vishwavidyalaya, Raipur in partial fulfillment of the requirements for the degree of M.Sc.(Ag.) in the department of Agricultural extension has been approved by the external examiner and students advisory committee after oral examination.

Date: 25/8/2014

External Examiner
(Dr. P.N. Sharma)

Major Advisor

Head of the Department

Dean Faculty

Director of Instructions

ACKNOWLEDGEMENT

"Education plays of fundamental role in personal and social development and teacher plays a fundamental role in imparting education. Teachers have crucial role in preparing young people not only to face the future with confidence but also to build up it with purpose and responsibility. There is no substitute for teacher pupil relationship".

From the care of my heart, I proudly avail this opportunity to express my warmest appreciation with deepest sense of gratitude to **Dr. P.K. Jaiswal** Professor, Department of Agricultural Extension, Rajmohani Devi College of Agriculture & Research Center, Ambikapur (C.G.) and Chairman of my Advisory Committee. I have no word to express my heartfelt thanks to him for invaluable inspiring guidance, unfailing encouragement, suggestions, research insight, unique supervision, constructive criticism, scholarly advice, sympathetic attitude and keen interest, throughout the investigation and preparation of this manuscript.

With great reverence, I express my sincere thanks to respected members of my advisory committee Dr. M.A. Khan College of Agriculture, IGKV, Raipur Dr. A.K. Sinha and Smt. Neelam Chouksey Rajmohani Devi College of Agriculture & Research Station, Ambikapur (C.G.) for their valuable suggestion and kind help rendered as and when necessary.

I express my sincere and profound gratitude to Dr. M.L. Sharma, Professor and Head, Department of Agricultural Extension, College of Agriculture, I.G.K.V., Raipur, whose inspiring suggestions, enthusiastic interest and encouragement provided me solace during the tenure of investigation and extending all kind of departmental help.

I wish to record my grateful thanks to Honourable Vice-chancellor Dr. S.K. Patil, Dr. J.S. Urkurkar, Director Research Services, Shri K.C. Paikara, Registrar, Dr. S.R. Patel, Dean, College of Agriculture, Raipur, Dr. S.S. Shaw, Director of Instruction, IGKV, Raipur, Dean, Rajmohani Devi College of Agriculture & Research Station, Ambikapur, Dr. R.B.S. Sangar and Dr. Dr. M.P Thakur, Director Extension Services for providing necessary facilities, technical and administrative supports for conductance of this research work.

I am highly obligated to all teaching staff members of Department of Agricultural Extension, Dr. J.D. Sarkar, Dr. K.K. Shrivastava, Dr. H.K. Awasthi, Dr. D.K. Surwanshi, Shri M.K. Chaturvadi and Shri P.K. Sagode for their co-operation, valuable suggestions and encouragement during the course of study. The staff members of Agricultural Extension Department deserve special thanks for their help and co-operation.

Words can hardly express the heartfelt gratitude to my beloved father Shri Rajendra Kumar Tiwari and mummy Smt. Shadhna Tiwari, whose selfless love, filial affection, obstinate sacrifices and blessing made my path easier. My most cordial thank goes to my younger Sister Smt. Divya Tiwari and my all family members. I give my special thanks to my best friends like Abhishek, Niharika, pradeep, Manish, Arivind, Dikshant, Nitin, Bhawesh Vikash, bhaskar, Surya, Ekta, pooja, lalit, Inder and koushik for their sweet memories which inspire me to work hard even in the critical period of this work.

I am highly thankful my seniors particularly Mr. Dilip Bande, Kedarnath yadav and Pandu Ram Paikra, for their earnest cooperation.

There are many friends and well wishers who helped me in various ways towards the present study and they deserve my sincere thanks. Among them my special thanks to Vinod Tiwari, Prashant, Dilip, gopal. I am also thankful to my juniors, Nagre, Parmeswar,Sonilal, Priyanka, Neesha, Shaivalini, Raja, Sohan and Kamal. I would like to convey my cordial thanks to all those who helped me directly or indirectly to fulfill my dream.

I would like to pray Almighty and Omnipresent "God" who is invisible but always with me as my mother and father for improving my confidence and determination in my whole life. So, my lord, please realize and accept my feelings.

Department of Agril. Extension.
College of agriculture,
I.G.K.V Raipur (C.G.)

RAHUL KUMAR TIWARI

CONTENTS

CHAPTER	PARTICULARS	PAGE NO.
I	**INTRODUCTION**	1-5
II	**REVIEW OF LITERATURE**	6-29
2.1	**Profile of sugarcane growers**	
2.1.1	Age	6-7
2.1.2	Farming experience	8
2.1.3	Education	8-10
2.1.4	Family size	10
2.1.5	Social participation	11
2.1.6	Occupation	12-13
2.1.7	Size of the land holding	13-14
2.1.8	Annual income	14-15
2.1.9	Credit acquisition	15-16
2.1.10	Scientific orientation	16-17
2.1.11	Contact with extension agencies	17-18
2.1.12	Source of information	18-19
2.1.13	Knowledge level	19-20
2.2	Adoption	20-24
2.3	Constraints	24-27
2.4	Suggestion	27-29
III	**RESEARCH METHODOLOGY**	30-46
3.1	Location of study	31
3.2	Sample and sampling procedure	31
3.3	Independent and dependent variables	32
3.4	Application of independent variables and their measurements	33
3.4.1	Independent variables	33-40
3.5	Application of dependent variable and its measurement	40
3.5.1	Extent of adoption of recommend sugarcane production technology	40
3.6	Constraints faced by the tribal sugarcane growers in adoption of recommended sugarcane production technology	41
3.7	Suggestion given by the tribal sugarcane growers to minimize the constraints	42
3.8	Type of data	42
3.9	Developing the interview schedule	**42**

3.9.1	Validity	43
3.9.2	Reliability	43
3.10	Method of data collection	44
3.11	Data processing and statistical frame work used for analysis of data	44-46
IV	**RESULTS AND DISCUSSION**	47-76
4.1	Independent variables	48
4.1.1	Age	48
4.1.2	Education	49
4.1.3	Family size	49-50
4.1.4	Size of the land holding	50
4.1.5	Occupancy of sugarcane land	51
4.1.6	Farming experience	51-52
4.1.7	Annual income	52
4.1.8	Credit acquisition	53-54
4.1.9	Occupation	55
4.1.10	Social participation	55-56
4.1.11	Scientific orientation	56
4.1.12	Contact with extension agencies	57-59
4.1.13	Source of information	59-60
4.1.14	Knowledge level	61-63
4.2	Dependent variables	63
4.2.1	Extent of adoption of recommended sugarcane production technology by tribal sugarcane growers	63-68
4.3	Correlation analysis of independent variables with adoption of recommended sugarcane production technology	68-69
4.4	Multiple regression analysis of independent variables with adoption of recommended sugarcane production technology	69-71
4.5	Constraints faced by the sugarcane growers in adoption of recommended sugarcane production technology	71-74
4.6	Suggestions offered by the tribal sugarcane growers to minimize the constraints faced by them	74-76
V	**SUMMARY, CONCLUSION AND SUGGESTIONS FOR FUTURE RESEARCH WORK**	77-88
	ABSTRACT	89-90
	REFERENCES	91-99
	APPENDIX	I-XX

LIST OF TABLES

TABLE NO.	PARTICULARS	PAGE NO.
4.1	Distribution of sugarcane growers according to their age	48
4.2	Distribution of sugarcane growers according to different categories of education level	49
4.3	Distribution of sugarcane growers according to size of family	49
4.4	Distribution of sugarcane growers according to their size of land holding	50
4.5	Distribution of sugarcane growers according to their different occupancy of sugarcane land holding	51
4.6	Distribution of sugarcane according to their farming experience	51
4.7	Distribution of sugarcane according to their annual income	52
4.8	Distribution of sugarcane growers according to their credit acquired	53
4.9	Distribution of respondents according to their occupation	55
4.10	Distribution of respondents according to their Social participation	55
4.11	Distribution of respondents according to their scientific orientation	56
4.12	Distribution of respondents according to overall contact with extension agency	57
4.13	Distribution of respondents according to their contact with individual extension agency	57
4.14	Distribution of respondent according to overall use of information source	59
4.15	Distribution of respondents according to their use of information sources	60
4.16	Distribution of respondent according to overall level of knowledge regarding recommended sugarcane production technology	61
4.17	Distribution of respondent according to their practices-wise level of knowledge regarding recommended sugarcane production technology	62
4.18	Distribution of respondents according to over all extent of adoption regarding recommended sugarcane production technology	64
4.19	Adoption level of sugarcane growers according to their different improved sugarcane cultivation practices	65
4.20	Correlation analysis of independent variables with extent of adoption of sugarcane production technology	68
4.21	Multiple regression analysis of independent variables with	70

		the adoption of recommended sugarcane production technology	
	4.22	Constraints faced by the sugarcane growers in adoption of recommended sugarcane production technology	72-73
	4.23	Suggestions of tribal sugarcane growers for minimizing the constraints faced by them during the adoption of recommended sugarcane production technology	75

LIST OF FIGURES

FIGURE NO.	PARTICULARS	BETWEEN PAGES
3.1	Map of the study area	30-31
4.1.1	Distribution of respondents according to their age	49-50
4.1.2	Distribution of respondents according to different categories of education level	49-50
4.1.3	Distribution of respondents according to their size of family	50-51
4.1.4	Distribution of respondents according to their land holding	50-51
4.1.5	Distribution of respondents according to their sugarcane farming experience	52-53
4.1.6	Distribution of respondents according to their annual income	52-53
4.1.7	Distribution of respondents according to their occupation	55-56
4.1.8	Distribution of respondents according to their social participation	55-56
4.1.9	Distribution of respondents according to their scientific orientation	57-58
4.1.10	Distribution of respondents according to over all contact with extension agency	57-58
4.1.11	Distribution of the respondents according to their contact with individual extension agency	57-58
4.1.12	Distribution of respondents according to their overall use of information source	60-61
4.1.13	Distribution of respondents according to their use of information source	60-61
4.1.14	Distribution of respondents according to overall knowledge regarding recommended sugarcane production technology	61-62
4.1.15	Distribution of respondents according to their practice wise level of knowledge regarding recommended sugarcane production technology	62-63
4.1.16	Distribution of respondents according to overall adoption regarding recommended sugarcane production technology	64-65
4.1.17	Distribution of respondents according to their practice wise level of adoption regarding recommended sugarcane production technology	65-66

Introduction

CHAPTER – I

INTRODUCTION

Sugarcane (*Saccharum officinarum* L.) is out of the important commercial crops of the world and is cultivated in about seventy five countries, The leading countries of which are India, Brazil, Cuba, Mexico and Thailand. The sugar industry plays an important role in the agricultural economy of India.

In terms of sugarcane production, India and Brazil are almost equally placed. In India, about 60 per cent of sugarcane is milled for the production of sugar, about 30 per cent for gur and khandsari, and the remaining 10 per cent is used for seed.

India occupies the second rank in production of sugarcane in the world. The area under sugarcane cultivation in India is 5.03 million hectares (2011-12) and cane production is 342.50 million tonnes with 68.09 tonnes/ ha productivity (*Directorate of Economics and Statistics, Department of Agriculture and Cooperation,* GOI, 2013). India's sugar production was estimated to be around 24.5 million tonnes as compared to annual consumption of 23 million tonnes.

Out of total 13.8 million ha geographical area of C.G. state, 45 per cent (63.49 Lakh hectares.) area is occupied by agriculture sector in the state of which 47.5 lakh ha (35% of geographical area) is net sown. The contribution of agriculture sector about 42 per cent in the state area economy. Over 80 per cent of the State population depends on agriculture and allied sectors for their livelihood. As per the agro-climatic zone, Chhattisgarh state falls under Eastern plateau and hill zone, which is further divided into three distinct sub-agro climatic zones *viz*, Chhattisgarh Plains, Northern Hills of Chhattisgarh and Bastar Plateau.

Out of total cultivated area in the state, about 74 percent area in Chhattisgarh plains, 95 per cent area in Northern hills and 97 per cent area in Baster Plateau is rain fed. All three zones received rainfall mainly through South-West mansoon during *kharif* and very few through North East mansoon during *rabi*. The average annual rainfall of the state is about 1350-1400 mm and the climate is characterized by hot-summers with maximum temperature reached up to 46 $^{\circ}$C during May- June and cold with minimum temperature up to 0 $^{\circ}$C during January in some part of the Northen hills of the state.

In Chhattisgarh, sugarcane is cultivated in 17.53 thousand hectares area with production and productivity of 45.42 thousand tons and 2591 kg/hectares, respectively during 2011-12. (Anonymous, 2013$_a$)

Surguja and Surajpur districts are important sugarcane growing districts of Chhattisgarh state. In 2011-12, the area occupies under sugarcane crop was 3.82 thousand hectare in both district (Surguja and Surajpur), production was 11.97 metric tonnes and productivity was 3134 kg per hectare. Lundra and Batauli, two blocks of Surguja district having sugarcane area of 1247.72 hectares and 550 hectares, respectively and Surajpur and Pratappur, two blocks of Surajpur district having 1077.79 hectares and 1507.63 hectares area, respectively during 2012-13.

The production and productivity of sugarcane can be increased by adoption of new technology as well as modernization in sugarcane cultivation practices as well as promoting farmers to grow sugarcane crop.

The sugarcane productivity has shown an increasing trend over the years. The magnitude has been quite trivial, wide gap exists between potential and the realized productivity. The gap between potential yield and realized yield may be due to the environmental factors, poor available improved varieties, low fertilizer application, poor pest and disease management, some socio-economic factors, inappropriate marketing facilities and post harvest related problems.

Although several crop improvement projects have implemented for the Indian agriculture sector by providing high yielding cultivars and production technologies in different states in case of sugarcane crop, it is a common observation that transfer of technology from lab to land is quite weak due to which many of these have not reached up to the ultimate users. This may be one of the reasons for poor sugarcane yield and sugar recovery as compared to potentiality of sugarcane yield and recovery. Sugarcane as well as sugar output can be increased, if the growers may motivated for adopting the recommended package of practices.

For increasing the level of adoption, farmers need to be convinced about recent knowledge about sugarcane production technology. In this regard, it is imperative to examine the existing status of knowledge and the factors influencing the process of adoption of new technology.

The present investigation was therefore carried out with the following objectives.

1. To study the socio - economic attributes of tribal sugarcane growers,
2. To ascertain the level of knowledge of the tribal sugarcane growers about recommended sugarcane production technology,
3. To find out the extent of adoption of sugarcane cultivation technology among selected tribal sugarcane growers,
4. To determine the relationship between characteristics of tribal sugarcane growers with extent of adoption regarding sugarcane production technology, and
5. To study the constraints perceived by the tribal sugarcane growers and obtain their suggestions to overcome them.

Significance of the study

The findings of this investigation will certainly help full to the extension system to redesign their activities for the transfer of technologies pertaining to sugarcane crop courdering the present state of production, productivity, marketing and socio-economic status of sugarcane growers. It will also help in addressing the major factors responsible for low production and productivity of sugarcane. It will also provide the specific feedback for the research system for technology assessment and refinement. The results will help the policy makers, administrators and planners associated with sugar industry and technology transfer sector of the state.

Limitations of the study

1. The investigation was conducted in a limited time with the restricted size of sample.
2. The method of data collection in the investigation was mostly confined to personal interview and the results drawn are exclusively based on the verbal expressed opinion and responses provided by the respondents.
3. Implication of the findings of the study will be applicable to the area of investigation and similar situation only.
4. The study is of limited geographical location Surguja and Surajpur districts. So the result may not lead to broader generalization.

Review of Literature

CHAPTER- II

REVIEW OF LITERATURE

One of the important aspects of research is the review of past literature. The researcher has to review the concerning literature at every stage. It is not a one shot exercise but a continuous process, while going through the literature, the researcher get acquainted with the subject-matter, techniques and materials and guide his efforts in desirable direction. Through review, researcher came to know about the methods, procedures and techniques as well as results of past studies. It provides clues and guidance throughout the research process. Steady efforts were made to compile research findings of the research studies possessing more or less similar characteristics. The present chapter incorporates all the relevant literature developed in India and abroad related to adoption of agricultural production technology of sugarcane by the farmers under following heads:

2.1. Profile of sugarcane growers

2.1.1 Age:

Tiwari and Lall (1998) found that age of farmers positively and significantly related with scientific attitude of sugarcane growers.

Kanavi (2000) in a study on the knowledge and adoption behaviour of sugarcane growers in Belgaum district of Karnataka state revealed that 42.66 percent of the respondents were in middle age followed by young age (30.66%) and old age (26.66%) respectively.

Nagaraja (2002) conducted a study on sugarcane growers in Davangere district of Karnataka, revealed that 78.33 per cent of the respondents were found with middle age category, whereas, 15.41 per cent and 6.25 per cent of respondents had fallen in the old and young age categories respectively.

Shrivastava *et al.* (2002) reported that age was positively and significantly related with the adoption level of chilly growers.

Naik (2005) conducted a study on knowledge and adoption pattern on improved sugarcane practices in Bidar district of Karnataka state concluded that 53.33 per cent of the respondents were in middle age followed by young age (13.34%) and old age (33.33%), respectively.

Maraddi (2006) conducted an analysis of sustainable cultivation practices followed by sugarcane growers in Karnataka revealed that 65 per cent of the respondents were middle aged (i.e., 36 to 50), whereas nearly 23 and 12 per cent of the respondents belonged to young age and old age groups, respectively.

Shivanand (2007) conducted HRD activities initiated by Nandi sugar factory Bijapur district of Karnataka state revealed that 46 per cent of the respondents were in middle age followed by old age (34.00%) and young age (20.00%) respectively.

Itawdiya (2011) in a study on technological gap in sugarcane production of sehore block of Madhya Pradesh revealed that most of the respondents (40.00%) were of middle age group followed by young age group (32.22%) and old age group (27.78%).

2.1.2 Farming Experience

Natikar (2001) in his study found that majority of the respondents belonged to medium farming experience (48%) followed by high (45%) and low (7%) farming experience, respectively.

Nagaraja (2002) in his study revealed that, a large majority of the respondents had four to eight years of experience in sugarcane cultivation as indicated by 85.41 per cent of the respondents, whereas, 13.85 per cent of the respondents mentioned 9 to 12 years as their sugarcane farming experience. Negligible per cent (0.83%) had the experience of more than 12 years in the sugarcane cultivation practices.

Maraddi (2006) revealed that medium experience (13 – 35 years) was possessed by 54.45 per cent sugarcane growers followed by low experience (<13 years) 24.44 per cent and high experience (>35 years) 21.11 per cent respectively.

Shivanand (2007) concluded that 38.67 per cent of the respondents were cultivating sugarcane for more than 8 years. About one third (32.66%) of the respondents had experience upto 4 years. Whereas, 28.67 per cent of the respondents had the experience of 5-8 years in sugarcane cultivation.

2.1.3 Education

Channal (1995) in a study on share holders and non-share holders of sugarcane growers in Belgaum district reported that 43.00 per cent of the share holders were under the category of primary and middle school followed by high school and pre university course (PUC) (36.00%), illiterate (17.00%) and graduate (4.00%), while in case of non-share holders 40.00 per cent had studied up to primary and middle school followed by high school and (PUC) pre University course (38.00%), graduates (12.00%) and illiterate (10.00%).

Kanavi (2000) found that 30.00 per cent of the sugarcane growers were illiterates followed by high school (22.00%), middle school (15.33%), primary school (11.33%), post graduates (9.33%) and 6 per cent in case of higher secondary and graduates.

Nagaraja (2002) in a study on sugarcane growers categorized the middle school (29.58%) and high school (30.83%), primary school (25.00%) whereas (8.33%) illiterate only 1.25 per cent of the respondents were graduate.

Naik (2005) found that 33.33 per cent of the sugarcane growers were primary school followed by High School (19.16), Middle School (17.50), College (12.52), illiterate (9.16) and Graduate were 8.33 per cent.

Maraddi (2006) found that 35.56 per cent of the sugarcane growers were educated up to primary school followed by middle (30.56%), illiterate (16.11%) and secondary (13.89%), whereas only 3.89 per cent of the respondents were educated up to collegiate level.

Shivanand (2007) found that 26 per cent of the respondents each were illiterate and had middle school education, followed by high school (18.00%), PUC (16.00%) and primary (12.00%) education.

Itawdiya (2011) concluded that (25.56%) of sugarcane growers were found to possess high school level education, and 24.44 per cent respondents were higher secondary passed, while 16.67 per cent respondents were educated up to college level. It is interesting to note that only 13.33 per cent respondents were illiterates whereas 8.89 and 11.11 per cent respondents possessed primary and middle school education respectively.

2.1.4 Family Size

Sharma (2001) concluded that type of family was positively and significantly associated with extent of adoption of rice and wheat recommended production technology.

Khan *et al.* (2002) reported that type of family had a non significant relationship with the adoption of organic farming practices by farmers.

Shrivastava *et. al.* (2002) observed that majority of chilli growers belonged to large family. i.e. more than 5 members.

Prajapati (2006) revealed that amongst selected wheat growers the majority of respondents (71.67%) belonged to joint family, whereas, only 28.33 per cent belonged to individual family.

Singh *et al.* (2007) observed that most of the farmers were living in joint family system in adopter (54.00%) and non adopter (51.00%), category of accelerating adoption of zero tillage system.

Suryawanshi (2009) observed that the majority of finger millet growers (63.33%) were residing in joint family system, however, 36.67 per cent of the respondents prefer to live in nuclear family system in the study area.

2.1.5 Social Participation

Tomar (1993) concluded that 71.66 per cent wheat growers had higher social participation followed by 38.34 per cent with lower social participation.

Sharma *et al.* (2000) found that social participation was significant and positively related to soyabean and maize cultivation practices.

Pal *et al.* (2001) observed that organization participation had significant relationship with the adoption of recommended practices of sugarcane cultivation.

Choudhary (2003) stated that the maximum number of the women respondents (34.54%) had no membership regarding in any organization scientific storage practices of food grains.

Dhruw (2008) concluded that the maximum (41.66%) number of the maize growers had membership in one organization.

Patel (2008) concluded in his study regarding soybean production technology that the 42.68 per cent respondents had no membership in any organization indicating very poor social participation.

Suryawanshi (2009) found that the maximum (41.33%) number of finger millets growers had membership in one organization.

Verma (2009) revealed that the maximum number (14.17%) of the paddy growers had membership in one organization followed by (6.67%) respondents had membership in more than one organization.

2.1.6 Occupation

Tomar (1993) concluded that the majority of the wheat growers (45.00%) were having agriculture occupation followed by 30.83 per cent agriculture + service.

Sharma *et al.* (2000) found occupation has the significant correlation with adoption level of crop production technology.

Pandey (2000) found that majority of the respondents (43.75%) were practicing only one occupation. Similarly about 37.50 per cent were engaged in two occupations and only limited i.e. 18.73 per cent were engaged in more than two occupation in relation to adoption of rice production technology.

Raghuvanshi (2003) reported that majority of the respondents (70.00%) were engaged in two occupation and only 1.25 per cent were practicing more than two occupation regarding adoption behaviour of rice growers.

Prajapati (2006) revealed that the majority of the wheat growers (61.67%) practiced only farming as the source of income, whereas 38.33 per cent farmers had other source of income along with farming.

Patel (2008) revealed that among the selected soybean growers maximum number of the respondents (52.00%) were involved in farming, followed by farming + labour (14.00%), farming + service (12.66%), farming + animal husbandry + service (7.34%) farming + others (8.00%) and farming + occupation + service (6.00%), respectively as their main occupation.

Kumar and Singh (2009) observed that the majority (93%) of the wheat growers were actively involved in agriculture for their livelihood.

2.1.7 Size of land holding

Kanavi (2000) categorized sugarcane growers in to large farmer (61.33%), medium farmers (30.66%), semi-medium farmers (6.55%) and small farmers (1.33%). None of the farmers belonged to category of marginal farmers.

Nagaraja (2002) in a study on sugarcane growers has classified the respondent as marginal farmer (2.50%), small farmer (2.08), semi-medium farmer (30.00%), followed by medium farmer (48.75%) and big farmer (16.66%).

Raghavendra (2004) conducted a study on knowledge of improved cultivation practices of sugarcane and their extent of adoption by farmers in Bhadra command area in Davanagere district, Karnataka and found that, majority of the respondents belonged to medium land holding (48.75%) followed by semi medium land holding category (30.00%).

Maraddi (2006) concluded that 45.55 per cent of the sugarcane growers belonged to the category of medium farmers followed by small farmers (22.77%) and semi-medium farmers (16.66%) and big farmers (15%). None of the farmers belonged to the category of marginal farmers.

Shivanand (2007) concluded that 32 per cent of the sugarcane cultivators belonged to the category of large farmers followed by medium farmers (30%), semi medium farmers (22%), small farmers (14%) and only (2%), marginal farmers.

Itawdiya (2011) concluded that out of the total 90 respondents, higher percentage (30.00%) of the respondents had medium size of land holding followed by 27.78%, 22.22% and 20.00% per cent respondents had large, small and marginal size of land holding respectively.

2.1.8 Annual income

Kanavi (2000) found that none of the sugarcane farmers were earning less than Rs.11,000/- per annum. The annual income of 2.00 per cent and 9.33 per cent of the respondents family was between Rs.11,000/- to Rs.22,000/- and Rs.22,000/- to Rs.33,000/- respectively.

Palaniswamy and Sriram (2001) observed in their study on modernization characteristics of sugarcane growers that, majority of the respondents (93.20%) belonged to high annual income group, while 4.76, 2.04 per cent of respondents belonged to medium and low annual income group, respectively.

Nagaraja (2002) in a study of sugarcane growers in Davangere district of Karnataka state revealed that 7.50 per cent of farmer having income of Rs.20,000, 48.33 per cent and 35.83 per cent of respondents having annual income of Rs.20,001/- to Rs.50,000/- and Rs.50,001 to 1 lakh respectively.

Naik (2005) concluded that 61.66%, sugarcane growers to be medium with income of i.e., Rs.68,590 to Rs.1,61,820/-, followed by low *i.e.,* 20.83 per cent respondents who have income less than Rs.68,590/-, and high *i.e.,* 17.5 per cent respondents have income more than Rs.1,61,820/-.

Itawdiaya (2011) found that out of the total sugarcane growers a similar higher percentage of the respondents i.e. 34.44 per cent earned both low and medium annual income while 31.12 per cent were earning high annual income.

2.1.9 Credit acquisition

Limje (2000) observed significant and positive correlation between the credit facilities and the adoption of soybean production technology.

Mukim (2004) indicated that the majority of respondents (96.09%) acquired the credit and credit acquisition had positive and significant association with the adoption of sunflower production technology.

Pandey et al. (2004) revealed that majority of respondents (62.96%) had taken the short-term credit followed by mid (25.92%) and long-term credit (11.12%) for rice crop.

Mishra (2006) observed that credit acquisition had a non-significant relationship with extent of adoption of recommended sugarcane production technology.

Prajapati (2006) revealed that the majority of the respondents (60.83%) had no credit facilities, whereas, 39.17 per cent of the respondents availed credit facilities for wheat production technology.

Dhruw (2008) indicated that the majority of the respondents (50%) had taken loan from nationalized bank for adoption of maize production technology.

Lanjewar (2009) concluded in his study regarding adoption of recommended cabbage production technology that majority of the respondents (55.00%) had acquired short term credit; cooperative society was the major source of credit and the credit facilities were available to them very easily and quickly.

2.1.10 Scientific orientation

Palaniswamy and Sriram (2001) observed in their study on modernization characteristics of sugarcane growers that, 70.75 per cent of respondents belonged to medium level of scientific orientation category. Whereas, 17.01 and 12.24 per cent of respondents belonged to high and low level of scientific orientation category, respectively.

Nagaraja (2002) in his study stated that, majority (67.08%) of the respondents had medium level of scientific orientation respect of improved package of agril. practices. The high level scientific orientation was seen in 22.08 per cent of the respondents. Whereas, only 10.83 per cent of the respondents had low level of scientific orientation.

Maraddi (2006) concluded that 46.11 per cent farmers were found in medium scientific orientation category followed by 35.56 per cent in low category and 18.33 per cent in high category.

Shivanand (2007) concluded that 68.0 per cent of sugarcane growers strongly agree to the statement "new method of farming gives better result to a farmer than the old methods" followed by 18.00 per cent of respondents who disagree whereas, only 6.00 per cent, 6.00 per cent and 2.00 per cent of respondents expressed that they agree, undecided and strongly disagree, respectively. A considerable percentage (42.00%) of the respondents are undecided to 'the way our fathers fore farmed' is still the best way of farming today followed by 32.00 per cent of respondents who strongly agree, whereas, 24.00 per cent and 2 per cent of respondents agree and disagree, respectively. Over half (58.00%) of the respondents strongly agree to 'even a farmer with lot of experience should use new methods of farming' followed by 14.00 per cent each agree and undecided, whereas, 12.00 per cent and of respondents disagree and strongly disagree, respectively. Nearly two-third (62.00%) of the respondents strongly agree to 'though it takes time for a farmer to learn new methods in farming it is worth the efforts', followed by 12.00 per cent each who agree and undecided whereas 14.00 per cent of the respondents disagree. 'A good farmer experiments with new ideas in farming', for this statement 52.00 per cent of the respondents strongly agree followed by 18.00 per cent of respondents who agree, whereas, 14.00 per cent each of respondents are undecided and disagree. Over half (56.00%) of the respondents strongly agree to 'traditional methods of farming have to be changed in order to raise the level of living of a farmer' followed by 16.00 per cent of respondents who are undecided and 14.00 per cent each of respondents agree and disagree.

2.1.11 Contact with extension agencies

Pandey (1996) found that majority of sugarcane growers 80.00 per cent had moderate level of extension contact in relation to entrepreneurial behavior of sugarcane growers.

Belligeri (1996) observed that 53 and 47 per cent had regular and occasional contacts with Agricultural Assistant. About 54 per cent of the respondents had occasional contact with Assistant Agriculture Officer, followed by regular contact (41.00%) and 5 per cent never contacted.

Tailor *et al.* (1998) reported that the contact with the extension personnel of both the categories of farmers (small and big) had positive relationship with the knowledge and adoption of selected dryland farming practices.

Mazher *et al.* (2003) found that extension agents communicated sugarcane technologies to 100 per cent of the farmers and a large majority of them adopted some of the technologies.

2.1.12 Source of information

Gupta *et al.* (2003) indicated that electronic media, television and radio are mostly used and preferred for seeking information on agriculture and development.

Rajni (2006) revealed that out of 14 selected information source, the training/visit system was utilized by cent per cent of the respondents. Friends (66.66%), neighbour (57.14%) and relatives (56.35 %), were utilized as a source of information. Whereas, equal percentage (54.76%) of respondents utilized television and news paper as source of information. Other information sources were agriculture magazine (37.30%), Sarpanch (30.95%) and only 12.69 per cent respondents, utilized village leader as source of information regarding mushroom production and processing training on farm women.

Deshmukh *et al.* (2007) observed that the majority of the respondents fall under medium utilization of sources of information (69.09%) category.

Singh and Kumar (2007) reported that all categories of the wheat cultivators used almost all the sources of information. However, majority of the farmers, irrespective of size of land holding had other farmers as their source of agriculture information. The large numbers were using, all the sources i.e. Press, T.V., Radio, other farmers, Agriculture Department and Research Institute to gather information. The trend was more or less similar in case of medium farmers too. The uses of press and T.V. were less amongst the small and marginal farmers. The farmers ranked fellow farmers first, followed by Agriculture Department, Research Institutes, Television, Radio and Press.

2.1.13 Knowledge level

Aski *et al.* (1997) in their study on, 'impact of training on knowledge and adoption pattern of sugarcane growers' noticed that, 18.57 per cent of trained respondents had 18.65 and above knowledge score, while 11.43 per cent of untrained respondents were found in high knowledge level category with 18.16 and above knowledge score.

Naik (2005) concluded that majority (60.83%) of the sugarcane farmers were found to belong to medium level of knowledge category. One fourth (25.84%) of the respondents had fallen in high knowledge category followed and 13.33 per cent of the respondents in to category.

Maraddi (2006) concluded that more than half of the sugarcane growers (53.33%) had medium knowledge level of sustainable cultivation practices, where low knowledge was possessed by 32.78 per cent of the respondents and in high knowledge category 13.89 per cent respondents were found.

Shivanand (2007) concluded that majority of the respondents (38.00%) had medium level of knowledge about improved production technologies of sugarcane with mean knowledge score of 14.35. Whereas, 32.00 per cent and 30.00 per cent of the respondents had low and high level of knowledge with mean score of 13.50 and 15.20, respectively.

Itawdiya (2011) concluded that majority of sugarcane growers (50.00%) had medium knowledge about sugarcane production technology followed by 25.56 per cent medium and 24.44 per cent had low knowledge about sugarcane production technology.

2.2 Adoption of improved sugarcane production technology

Kharde and Nimbalkar (1986) conducted a study on sugarcane growers and reported that 60 per cent sugarcane growers were in medium adoption category. While, 32.67 and 7.33 per cent were observed in low and high adoption categories, respectively.

Aski (1989) conducted a study in Belgaum district of Karnataka on trained and untrained farmers of Farmers Training Centre (FTC) on sugarcane cultivation and pointed out that 67.15 per cent of the trained and 65.71 per cent of untrained farmers belonged to medium adoption category. Only 15.17 per cent and 20 per cent of trained and untrained farmers fell under low adoption category.

Vekaria *et al.* (1990) in a study on role played by co-operative sugar factories in sugarcane development found that majority of the respondents (68.60%) who possessed favourable attitude towards co-operative sugar factory were medium adopters and the rest of them 18.18 and 13.22 per cent were high and low adopter, respectively. Whereas, the majority of the respondents (67.05) who possessed unfavourable attitude towards co-operative sugar factory were medium adopters and 13.63 and 19.32 per cent were high and low adopter, respectively.

Channal (1995) while studying the knowledge and adoption behaviour of shareholders and non-shareholders of cooperative sugarcane factory in Belgaum district of Karnataka reported that 37 per cent of shareholders and 36 per cent of non shareholders were in the medium adoption category. Thirty three per cent of shareholders and thirty five per cent of non-shareholders belonged to high adoption category. Remaining 30 per cent of the shareholders and 29 per cent non-shareholders fell under low adoption category.

Karthikeyan *et al.* (1996) in their study revealed that majority of the sugarcane growers (73.33%) were medium level adopters. One fifth of them belonged to high adopter's category.

Bhatkar *et al.* (1997) conducted a study in Akola districts of Maharashtra state, indicated that majority of the sugarcane growers (64.00%) were medium in adoption of recommended practices of sugarcane crop. Whereas, one fifth (20.67%) respondents were in high level and 15.33 per cent were low adopters.

Aski *et al.* (1997) in their study on impact of training on knowledge and adoption pattern of sugarcane growers found that, 17.14 per cent of trained farmers belonged to high adoption category, while 14.29 per cent of untrained were found in high adoption category.

Goud (1998) in a study on adoption levels of registered and non-registered growers in Sameerwadi sugar factory area of Bagalkot district, Karnataka, reported that 51.17 per cent of registered sugarcane grower adopted earthingup followed by spacing (40.05%), fertilizer application (33.33%), seed rate (26.92%), variety (25.06%), plant protection measures (14.67%), and ratoon management (14.61%). Whereas, in case of unregistered growers 34.33 per cent of sugarcane growers adopted earthingup followed by fertilizer application (20.87%), variety (16.11%), spacing (15.23%), seed rate (13.33%), ratoon management (94.78%) and plant protection measures (4.17%).

Krishnamurthy *et al.*, (1998) in their study conducted in Doddi sugar factory area of Mandya district of Karnataka reported that cent per cent of the respondents followed simple practices like sowing time, variety, seed rate and time of harvest, whereas, 80 per cent of farmers followed practices like quantity of FYM, application of potash and hand weeding.

Kanavi (2000) revealed that majority (78.00%) of the sugarcane growers belonged to medium adoption category, only 14.66 per cent and 7.33 per cent of the respondents belonged to low and high adoption category.

Nagaraja (2002) concluded that 61.24 per cent of the sugarcane growers were found in the medium level of adoption category. About one fourth (25.83%) and 12.91 per cent of the respondents had fallen in low and high adoption categories respectively.

Pawar *et al.* (2005) conducted a study was in western Maharashtra, India during 2002-03 to investigate the extent of adoption of improved technology for sugarcane growing among 270 growers in the 3 recovery zones of the region. The gross cropped area was 4.05 ha and the cropping pattern was dominated by sugarcane (57.28 %). Sugarcane production was highest in the high recovery zone (83.66 %), followed by the medium and low recovery zones (52.20 and 41.35 % respectively) the awareness of adoption of improved sugarcane production technology in terms of soil and plantation time was > 90%, the adoption of preparatory tillage was 66 % and the adoption of recommended fertilizer nutrients was 15% due to unawareness of requirements of soil testing based fertilizer use. Poor adoption of two -bud set and one-bud set and green manure was observed. More than 85 % of the farmers were aware of ratoon management practices and adoption was > 65 %. Awareness of wooly aphid's incidence was 90 % and adoption of control measures was 48%.

Kathiresan *et al.* (2003) observed that most farmers were aware of the technologies and that non-availability of labour was the reason for non-adoption in the case of detracting and propping.

Naik (2005) observed that majority (55.83%) of the Sugarcane growers were found in the medium level of adoption category followed by high (26.66%) and 17.51 per cent of the respondents fallen under low category.

Maraddi (2006) found that majority (63.33%) of the respondents belong to medium overall adopter category followed by low (26.67%) and 10 per cent of sugarcane growers belong to high adopter category.

Shivanand (2007) reported that 54.00 per cent of the respondents belonged to medium adoption category with mean adoption scores of 9.58. Nearly one fourth each (24.00% and 22.00%) of the respondents belonged to high and low adoption category with mean adoption score of 10.19 and 8.98, respectively.

Itawdiya (2011) concluded that most of respondents (45.56%) had high adoption about sugarcane production technology followed by 33.33 per cent low and 21.11 per cent had high adoption improved technology.

2.3 Constraints in the adoption of improved sugarcane production technology.

Agarwal and Goswami (1992) analysed the constraints in sugarcane cultivation in Uttar Pradesh and stated that the major constraints were non-availability of credit, method of price fixation, time gap between harvesting and crushing of sugarcane.

Singh *et al.* (1993) assessed the problems of sugarcane production in Bijnor District of Uttar Pradesh reported that delay in supply of credit by the cane cooperatives to the growers, dominance of local leaders in the operation of cane cooperatives, transportation problems and lack of suitable extension services regarding new technology of production and disposal of sugarcane were the major problems.

Ramsingh and Singh (1994) carried out a study on the sugarcane production in Uttar Pradesh and reported that non-availability of inputs, long distance to be travelled to purchase inputs, lack of knowledge in adopting inputs and high cost of inputs were the major problems faced by the sugarcane growers.

Channal (1995) conducted a study in Belgaum District of Karnataka and reported that 69 per cent of shareholders and 100 per cent of non-shareholders expressed delay in buying of cane by the factory as the main problem. Thirty one per cent of shareholders and 47 per cent of non-shareholders expressed bills were not paid in time. Fifteen per cent shareholders and 32 per cent non-shareholders expressed permit by the factory for cutting cane from factory was not given in time. Further, the general problems faced by the shareholders were power supply (61.00%), labour problem (48.00%) non availability of credit (22.00%) non availability of fertilizer in time (24.00%), water problem (18.00%) and high cost of fertilizer (4.00%) were the major problems in cultivation of sugarcane.

Bhatkar *et al.* (1996) in a study conducted in Akola district of Maharashtra reported that, lack of knowledge about chemical weed control, fertilizer management, non availability of organic manure, lack of knowledge about planting of cane sets and proper position of eye buds, less price offered by sugar factory and lack of irrigation potential to cover more area under sugarcane cultivation were the major problems.

Karthikeyan *et al.* (1996) conducted a study on sugarcane growers of Pondicherry Co-operative Sugar Mills Limited (PCSM), Pondicherry. They found that more than one third of Tamil Nadu farmers faced the problems like belated payment for the produce (38.33%) and delayed disbursement of sanctioned loans (36.66%), further, 54.16 per cent of respondents expressed high cost of inputs followed by high labour cost (49.16%), low price to the produce (47.49%), high cost of cultivation (3.33%) and high transport cost (30.00%).

Krishnamurthy et al. (1998) in their study conducted in Mandya district of Karnataka revealed that majority (60.00%) of the farmers expressed lack of technical knowledge, non availability of inputs like credit, fertilizers and pesticides were the main reasons for non adoption of recommended practices of sugarcane cultivation.

Kanavi (2002) conducted a study in North Karnataka and revealed that, 67.33 per cent of the sugarcane farmers expressed irregular supply of electricity for irrigation followed by high cost of fertilizers and chemicals (54.66%), delay in cutting order and payment (53.33%), non availability of credit on time (34.66%), non availability of labour and their high wages (29.33%), low price to produce (25.33%), high commission charges by middle man (20.66%), non availability of inputs (11.33%) and drying up of open wells in summer.

Nagaraja (2002) in his study reported that, the major constraints faced by more than 90.00 per cent of sugarcane growers were high cost of fertilizers (92.48%) followed by delay in release of crop loan by the respective banks (96.65%), however, 80.82 per cent of the respondents faced the problem of high interest rate for crop loan. Around three fourth (76.33%) of the respondents expressed the problem of delay in transport of harvested cane from field to the factory followed by delay in issuing permit by the factory (76.23%) and no timely and proper technical guidance by the extension workers in the advanced technologies of sugarcane cultivation (72.90%). However, one-tenth (9.99%) of the respondents faced the problem of power supply in summer season.

Shivanand (2007) concluded that over three fourth of the respondents (80.67%) and 71.33 per cent of respondents expressed non availability of labour and irregular supply of electricity for irrigation, respectively. Other problems expressed by the respondents were price problem at factory (30.67%), technical problems in production (27.33%) followed by payment problem (12.00%).

2.4 Suggestion of sugarcane growers

Nagaraja (2002) in his study revealed that, majority (68.73%) of the sugarcane growers felt the need for training in the area of ratoon crop cultivation practices (68.73%) and also in the management of NPK nutrients in sugarcane cultivation (66.23%). More than half (56.24%) of the respondents interviewed, expressed the need for training in the area of package of practices in growing recent high yielding varieties followed by plant protection measures to be practiced in sugarcane cultivation (47.49%). It was also observed that, 22.07 per cent of the respondents felt the need for training in the area of use of micronutrients in sugarcane cultivation.

Maraddi (2006) concluded the suggestions expressed by sugarcane growers. In light of constraints faced, the major suggestion expressed was providing credit at lower interest rate and at required time, cost of complex fertilizers should be reduced, conduct demonstrations on different sustainable cultivation practices in sugarcane to show their efficacy, proper schedule of varietal harvesting must be followed by sugar factories, organize training programmes on sustainable cultivation practices in sugarcane, provide pest and disease resistant varieties through sugar factories and research stations/KVKs, conduct as many as group discussions, krishimela and exposure trip to sugarcane growers to convince the benefit of various sustainable cultivation practices, establish sugarcane growers club and conduct regular meetings with scientist and progressive farmers.

Shivanand (2007) concluded that about one third of the respondents (25.33%) expressed the suggestion that 'adult education for illiterates and higher education for children', followed by 22.67 per cent of the respondents expressed the suggestion that 'factory should award to best sugarcane growers'. 14.67 per cent, 10.67 per cent and 10.00 per cent of the respondents expressed suggestions like 'factory should provide incentives for shareholders', 'communication facilities' and 'special extension programmers, respectively. Whereas, 7.33 per cent, 5.33 per cent and 4.00 per cent of the respondents expressed the suggestions such as improvement of health campaigns, good transport facilities and crop competition respectively.

Itawdiya (2011) concluded that majority of the respondents (68.89%) has suggested that electricity supply should be regular at the time of irrigation, ranked as first major suggestion followed by 61.11 per cent of respondent suggested that market information about sugarcane should be provided ranked as second and 56.67 per cent suggested that Government should increase subsidies for purchase of fertilizers and other production inputs as third. Whereas, 52.22, 51.11, 50.00, 47.78, and 46.67 per cent of the respondents suggested that to develop comprehensive package for management of pests and diseases of sugarcane, sugarcane should be included under crop insurance scheme of Government of India, adequate credit facilities should be availed at farmer's door step, to develop locale-specific pest and disease resistant varieties of sugarcane and there is an urgent need for regulated market for sugarcane ranked as fourth, fifth, sixth, seventh and eighth respectively.

Research Methodology

CHAPTER - III

RESEARCH METHODOLOGY

The study "A STUDY ON TRIBAL FARMERS OF SURGUJA DIVISION WITH REFERENCE TO ADOPTION OF SUGARCANE PRODUCTION TECHNOLOGY" conducted during 2013-14 in four blocks of Surguja and Surajpur districts. A details account of material used, survey procedure followed and techniques adopted during the course of investigation are presented in this chapter under the following heads:

3.1 Location of study

3.2 Sample and sampling procedure

3.3 Independent and dependent variables

3.4 Operationalization of independent variables and their measurement

3.5 Operationalization of dependent variable and their measurement

3.6 Constraints faced by the tribal sugarcane growers in adoption of recommended sugarcane production technology

3.7 Suggestions given by the tribal sugarcane growers to minimize the constraints

3.8 Type of data

3.9 Developing the interview schedule

3.10 Method of data collection

3.11 Data processing and statistical framework used for analysis of data

3.1 Location of study

The study was conducted in Surguja and Surajpur district of Chhattisgarh state during the year 2013-14. Surguja and Surajpur district is situated at a distance of 350 km away from Indira Gandhi Krishi Vishwavidyalaya, Raipur. Surguja and Surajpur district is situated in Northern hill zone of Chhattisgarh state.

3.2 Sample and sampling procedure

3.2.1 Selection of district

Surguja+ Surajpur district is the 2nd largest producer of noble cane and the area (3.82 Thousand ha.) under cultivation is high in Chhattisgarh, therefore it was purposively selected for the present study

3.2.2 Selection of block

Out of 13 blocks of both districts, four blocks have been selected randomly for study, namely Pratappur (1507.63 ha.), Surajpur (1077.79 ha.) from Surajpur district and Lundra (1634 ha.), Batauli (1634 ha.) from Surguja district during 2012-13 respectively (DDA office, 2012-13).

3.2.3 Selection of village

Two villages from each selected block have been selected randomly for study. Thus survey as per objective of study work made in 8 villages namely Batwahi and Mahora (Lundra block), Mangari and Sarmana (Batauli block),

Haripur and Kalyanpur (Surajpur block) and Kerta and Khadgawakala (Pratappur block).

3.2.4 Selection of respondents

A list of tribal sugarcane growing farmers was prepared who were cultivating sugarcane from last three years, with the help of RAEOs of the eight villages. Sixteen tribal sugarcane growers have been selected randomly from each of the selected village. Thus the total 128 Sugarcane growers (16X8) =128 was considered as respondent for this study.

3.3 Independent and dependent variables

3.3.1 Independent variables

S. No.	Variables	Measurement tools
	Independent variables	
1.	Age	As per Chronological age
2.	Education	Pareek and Trivedi (1963)
3.	Sugarcane growing experience	In years
4.	Family size	In year
5.	Social participation	
6.	Occupation	Siddaramaiah and Jalihal (1983)
7.	Size of land holding	In ha.
8.	Annual Income	Self scoring
9.	Credit acquisition	
10.	Knowledge of practices	Self scoring
11.	Scientific orientation	Supe (1975)
12.	Contact with extension agencies	Self scoring

3.3.2 Dependent variables

❖ Adoption of recommended sugarcane production technology.

3.4 Application of independent and their measurement

3.4.1 Independent variables

(1) Age

The personal interview was conducted among respondents of different age group during course of study and was arranged as per chronological age. The data were categorized as follows:

S. No.	Categories	Scores
1.	Young age group	(20 to 35 years)
2.	Middle age group	(36 to 55 years)
3.	Old age group	(Above 56 years)

(2) Farming experience

It refers the experience of a farmers of sugarcane cultivation either on his own farm or leased farm. As per the experiences, scores has been given to sugarcane growers. The respondents were classified as low, medium and high experienced groups-

S. No.	Categories	Scores
1.	Low	(Up to 5 years)
2.	Medium	(6 to 10 years)
3.	High	(More than 10 years)

(3) Education

The reading and writing ability of respondents were considered as their education status and it was categorized as under:-

S.No.	Education	Scores
1.	Illiterate	0
2.	Primary school (1st to 5th)	1
3.	Middle school (6th to 8th)	2
4.	High School (9th to 10th)	3
5.	Higher Secondary School (11th to 12th)	4
6.	College and above	5

(4) Size of family

It reflects the number of members live together with respondents. They were categorized as follows:-

S.NO.	Categories	Score
1.	Small (1-5 Member)	1
2.	Medium (6-10 Member)	2
3.	Large (More than 10 Member)	3

(5) Social participation

The term social participation in this study refers to the degree of involvement of the respondents in formal / informal organization as a member or executive, office bearer or both. A social participation score was computed for each respondents on the basis of his membership and position in various formal informal organizations and categorized into following subheads:

S.NO.	Categories	Score
1.	No member in any organization	0
2.	Member of one organization	1
3.	Member of more than one organization	2
4.	Office bearer in any organization	3

(6) Occupation

The data collected from the respondents about their occupation were categorized into 6 groups and measured with score assigned as:-

S.NO.	Categories	Score
1.	Farming	1
2.	Farming + Labour	2
3.	Farming + Service	3
4.	Farming + Animal Husbandry	4
5.	Farming + Business	5
6.	Farming + Other	6

(7) Size of land holding

It was operationally defined as the actual land holding of the respondents at the time of investigation. The categorization of the respondents was done under the following subheads:

S. No.	Categories	Scores
1.	Marginal land holing (up to 1 ha.)	1
2.	Small land holding (1 to 2 ha.)	2
3.	Medium land holding (2 to 4 ha.)	3
4.	Large land holding (>4 ha.)	4

(8)　Annual income

It refers to the income earned in rupees by the respondents from agriculture and other occupation. Based on the total annual income, the respondents were categorized in to three groups:

S.No	Category	Scores
1	Low	(Up to Rs. 27,231)
2	Medium	(Rs. 27,232 to 1,99,324)
3	High	(> 1,99,324 Rs.)

(9)　Credit acquisition

The availability of credit needed to purchase the required inputs may influence the extent of adoption of new techniques to the farmers. The adoption of improved agricultural technology requires more investment of capital in farming to purchase the inputs like fertilizers, pesticides, improved seed, implements etc. Sources of credit were identified including cooperative society, nationalized banks, moneylenders, friends, neighbours, relatives, etc and each source was given equal weight-age and the availability of credit identified by farmers were then measured by the following scores.

Categories	Score
Not acquired	1
Acquired	2
Period of credit	
Short-term	1
Medium-term	2
Long-term	3
Availability of credit	
Acquired easily	1
Acquired with difficulty	2

(10) Level of knowledge about recommended sugarcane production technology

Knowledge about innovation may be an important factor affecting the adoption behavior of farmers. Bloom defined knowledge as "those behavior and best situation which emphasized the remembering either by recognition or recall of ideas, materials on phenomenon." Operationally knowledge was used in this study as actual knowledge of farmers regarding four selected practices i.e. seed rate, sowing time, fertilizer dose etc.

The knowledge test composed of items called questions for constructing the pertaining the package of practices of sugarcane production technology. The set the question developed was discussed with the subject matter specialist in the disciplines of advisory committee and finalized.

A device was developed to measure the knowledge level of farmers regarding selected technologies recommended for sugarcane crop, and was used with some modifications. The responses of respondents regarding knowledge were obtained into three point continuum as under.

S.No.	Categories	Score
1.	Partially level knowledge	1
2.	Medium level knowledge	2
3.	Complete level knowledge	3

$$K.I. = Mean\ (X) \pm S.D.\ (Standard\ Deviation)$$

Categories	
Low level of knowledge	($< \bar{X} - S.D.$)
Medium level of knowledge	(In between $\bar{X} \pm S.D.$)
High level of knowledge	($> \bar{X} + S.D.$)

(11) Scientific orientation

It refers to the degree to which an individual is inclined to use scientific method in farming and decision-making. The scientific orientation scale developed by Supe (1975) was used for the measurement of scientific orientation of respondents. The statements of the original scale were suitably modified to measure the scientific orientation of respondents. The scale has six items. Out of these six items, number 1, 3, 4, 5 and 6 were positive items and number 2 was a negative item. The score for positive item were 5, 4, 3, 2 and 1 and for negative item scores were 1, 2, 3, 4, 5 for the response categories strongly agree, agree, undecided, disagree and strongly disagree, respectively. The sums of scores of all the six statement were worked out. The respondents were categorized into following groups on the basis of following formula.

SOI = Mean $(\overline{X})$ ± S.D. (Standard Deviation)

Categories	
Low level of scientific orientation	($<\overline{X}$ - SD)
Medium level of scientific orientation	(in between $\overline{X}$ ± SD)
High level of scientific orientation	($>\overline{X}$ + SD)

(12)　Contact with extension agencies

This is operationally defined as the "frequency with which a respondents comes in contact with extension agents i.e. RAEOs, ADOs, SADOs, Subject matter specialist (SMS) and Agriculture scientist within a specific period of time". The extent of contact was measured into four categories viz., never, twice or thrice in a year, once in a month and weekly with score 0, 1, 2 and 3 respectively.

Ex. contacts = Mean (X) ± S.D. (Standard Deviation)

Categories	
Low level of extension contact	($<\overline{X}$ - SD)
Medium level of extension contact	(in between $\overline{X}$ ± SD)
High level of extension contact	($>\overline{X}$ + SD)

(13) Sources of information

A set of 13 information sources were identified including personal, group and mass media etc and each source was given equal weight-age and categories were made according to the use of information sources.

The respondents were grouped in to three categories for use of information sources by using following formula:

S.O.I. = Mean ($\bar{X}$) ± S.D. (Standard Deviation)

Categories	
Low level of use of information sources	($<\bar{X}$ − S.D.)
Medium level of use of information sources	(in between $\bar{X}$ ± S.D.)
High level of use of information sources	($>\bar{X}$ + S.D.)

The responses of respondents regarding use of information sources were obtained into three point continuum as under.

Categories	Score
No use	1
Occasional use	2
Frequent use	3

3.5 Operationalization of dependent variable and its measurement

3.5.1 Extent of adoption of recommended sugarcane production technology

It is mental process through which an individual passes from hearing about an innovation to final adoption (Roger, 1995).

It is operationalized as the degree of the use of recommended practices. Adoption refers to the extent of use of recommended farming practices of sugarcane cultivation by sugarcane growers. Extent of adoption respondents about practices in sugarcane cultivation was measured by undertaken the recommended package of practice's for higher yielding production of sugarcane released in the year 2010 by Indira Gandhi Krishi Vishwavidyalaya, Raipur.

To measure extend of adoption, recommended important practices were listed and responses for the each practices were obtained into three-point scale as under.

S.No.	Categories	Score
1.	Partially Adopted	1
2.	Medium Adopted	2
3.	Full Adopted (complete)	3

The respondents were classified into three categories by using following formula:

A.I. = Mean ($\bar{X}$) ± S.D. (Standard Deviation)

Categories	
Low level of adoption	($<\bar{X} - S.D.$)
Medium level of adoption	(In between $\bar{X} \pm S.D.$)
High level of adoption	($>\bar{X} + S.D.$)

3.6 Constraints faced by the tribal sugarcane growers in adoption of recommended sugarcane production technology

To measure the constraints faced by the tribal sugarcane growers in adoption of recommended sugarcane production technology, simple ranking technique was applied, each farmer was asked to mention his constraints in adoption of recommended sugarcane production technology in order of degree of difficulties.

3.7 Suggestions given by the tribal sugarcane growers to minimize the constraints

Considering the constraints faced by the tribal sugarcane growers regarding adoption of sugarcane production technology and to overcome the same in adoption of sugarcane production technology successfully, farmers were asked to give their valuable suggestions. The suggestions offered were summed up and ranked on the basis of number and per cent of farmers who reported for the respective suggestions.

3.8 Type of data

The data pertaining to selected characteristics about socio-personal, socio-economic, technological, communicational, socio psychological, adoption, constraints perceived in terms of adoption and suggestions of respondents were collected as per objectives of the study as primary data. The official information and records were also consulted from the concerning departments as secondary data.

3.9 Developing the interview schedule

The interview schedule was designed on the basis of objectives independent and dependent variables in the present investigation. To facilitate the respondents, the interview schedule was framed in "Hindi". Each question was thoroughly examined and discussed with the experts before presenting the interview schedule. Adequate precautions and care were taken into consideration to formulate the questions in a manner that they were well understood by the respondents and would find it easier to respond.

The prepared interview schedule was used in the study area for collecting the data. On the basis of experience gained in pre-testing, the necessary modifications and suggestions were incorporated before giving a final touch to interview schedule.

3.9.1 Validity

Validity refers to "the degree to which the data collection instruments measures what it is supposed to measure rather than something else". Taking the following steps validity of interview schedule used for this study was maximized:

1. The interview schedule was thoroughly discussed with the scientists and their suggestions were incorporated.
2. Pre-testing of interview schedule provided an additional check for improving the instruments.
3. The relevancy of each question in terms of objectives of study, logical order and wording of each question were checked carefully.

3.9.2 Reliability

Reliability of an interview schedule refers to "its consistency or stability in obtaining information from respondents".

The test-retest method of estimating reliability of an interview schedule was followed in this study. Twenty respondents of the study area were randomly selected and were re-interviewed after 2 to 3 weeks using the same interview schedule followed at the time of first interview. Since same responses were observed, the reliability of the interview schedule was ensured.

3.10 Method of data collection

Respondents were interviewed through personal interview. Prior to interview, respondents were taken into confidence by revealing the actual purpose of the study and also full care was taken to develop good rapport with them. They were assured that the information given by them would be kept confidential. The interview was conducted in the most informal and friendly atmosphere without any complications.

3.11 Data processing and statistical framework used for analysis

The data collected during the course of investigation was tabulated into the coding sheet and then appropriate analysis of data was made according to objectives as suggested by Cochran and Cox (1957). The statistics applied were percentage, frequency, mean, standard deviation, coefficient of correlation, multiple regression etc. the analysis was carried out with help of Computer Section of IGKV, Raipur.

3.11.1 Frequency and percentage

Frequency and percentage were used for making simple comparison.

3.11.2 Mean and standard deviation

3.11.2.1 Mean

Mean of sample was calculated by using the following formula:

$$\overline{x} = \frac{\sum x}{n}$$

Where,

$\overline{x}$ = Mean of the variable

$\sum x$ = Sum of score (observation) of variable

n = Total number of respondents

3.11.2.2 Standard deviation

Standard deviation was calculated by using following formula:

$$SD = \sqrt{\frac{1}{(n-1)}\left[\frac{\sum x^2 - (\sum x)^2}{n}\right]}$$

Where,

SD = Standard deviation

$x = x_i - \overline{x}$ = Deviation obtained from mean

n = Number of observations

3.11.3 Pearson's coefficient of correlation

This technique was used to find out the relationship between two variables. The formula used was as follows:

$$r = \frac{n\Sigma xy - \Sigma x \Sigma y}{\sqrt{n\Sigma x^2 - (\Sigma x)^2} \cdot \sqrt{n\Sigma y^2 - (\Sigma y)^2}}$$

Where,

    r = Correlation coefficient

    x = Score of independent variable

    y = Score of dependent variable

    n = Number of observation

3.11.4 Multiple regressions

This technique was used to know the partial and complete influence of independent variables. For the present study linear model of regression equation was used which is as follows:

$$Y_1 = a + b_1x_1 + b_2x_2 + \ldots\ldots\ldots + b_nx_n$$

Where,

Y_1 = Dependent variable

    $x_1\ldots x_n$ = Independent variables

    a = Constant value

$b_1\ldots b_n$ = The regression coefficient for respective independent variables.

Results and Discussion

CHAPTER-IV

RESULTS AND DISCUSSION

In this chapter, an attempt has been made to present the results and discussion during course of investigation. The information collected from sugarcane growers (128) through the structural interview have been summarized in relevant tables after statistical analysis keeping in view the specific objectives of the study. The observed results as supported by the available statistics have been described to project a broad and clear variation exhibited by sugarcane production technology.

The findings of the study are presented and discussed under the following heads:

4.1   Profile of tribal sugarcane growers/ Independent variables

4.2   Dependent variables

4.2.1 Adoption of recommended sugarcane production technology by tribal sugarcane growers.

4.3   Correlation analysis of independent variables with adoption of recommended sugarcane production technology by tribal sugarcane growers.

4.4   Multiple regressions of independent variables with adoption of recommended sugarcane production technology.

4.5 Constraints faced by the tribal sugarcane growers during adoption of recommended sugarcane production technology.

4.6 Suggestions made by the tribal sugarcane growers to overcome the constraints faced during adoption of recommended sugarcane production technology.

4.1 Profile of tribal sugarcane growers
4.1.1. Age:

Table 4.1 Distribution of sugarcane growers according to their age

(n=128)

S.No.	Particulars	Frequency	Per cent
1.	Young(<35) Years	30	23.43
2.	Middle(36 to 55 Years)	85	66.42
3.	Old (>55 Years)	13	10.15
	Total	128	100.00

The data pertaining to age have been presented in table 4.1 It is clear from data that maximum respondents belong to middle age group (66.42%). The remaining respondents belong under the categories of young age and old age group 23.43 per cent and their per cent observed as 10.15 per cent, respectively.

Perusal of data presented in table 4.1 revealed that majority of the tribal sugarcane farmers were under middle age group (36 to 55 years). Similar findings have also been reported by Kanavi (2000), Nagaraja (2002), Shrivastava *et al.* (2003), Naik (2005), Maraddi (2006), Kumar and Singh (2009), Itawdiya (2011) and Chauhan *et al.*(2013).

4.1.2. Education:

Table 4.2: Distribution of sugarcane growers according to different categories of education level. (n=128)

S.No.	Particulars	Frequency	Per cent
1.	Illiterate	28	21.87
2.	Primary	27	21.09
3.	Middle	33	25.78
4.	High school	13	10.15
5.	Higher secondary school	22	17.18
6.	College and above	5	3.93
	Total	128	100.00

The data presented in table 4.2 regarding education of respondents showed that maximum respondents had middle school level of education (25.78%) followed by primary school (21.09%), illiterate (21.87%), higher secondary school (17.18%), high school (10.15%) and only 3.93 per cent of the respondents had college and above education level. Channel (1995), Nagaraja (2002), Dhruw (2008), Chauhan et al, (2013) also reported similar findings.

4.1.3. Size of family

Table 4.3: Distribution of sugarcane growers according to size of family. (n=128)

S.No.	Particulars	Frequency	Percent
1.	Small (up to 5)	81	63.28
2.	Medium (6 to 10)	42	32.81
3.	Big (>10)	5	3.91
	Total	128	100.00

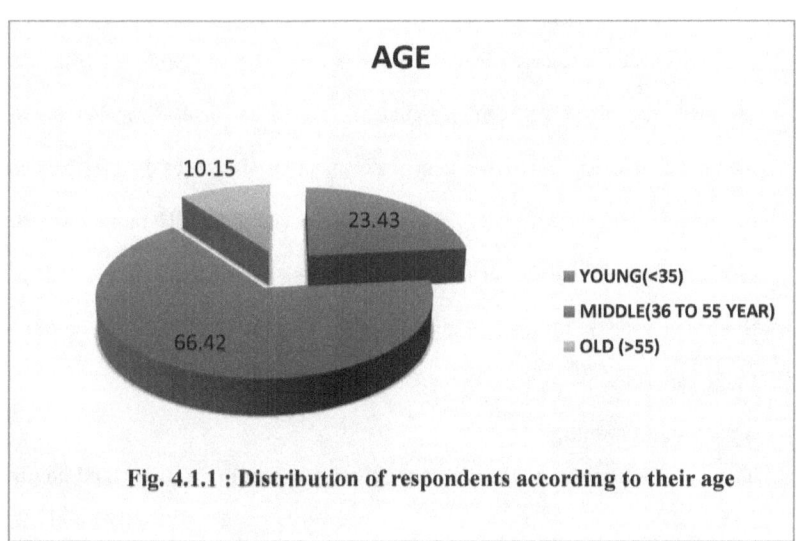

Fig. 4.1.1 : Distribution of respondents according to their age

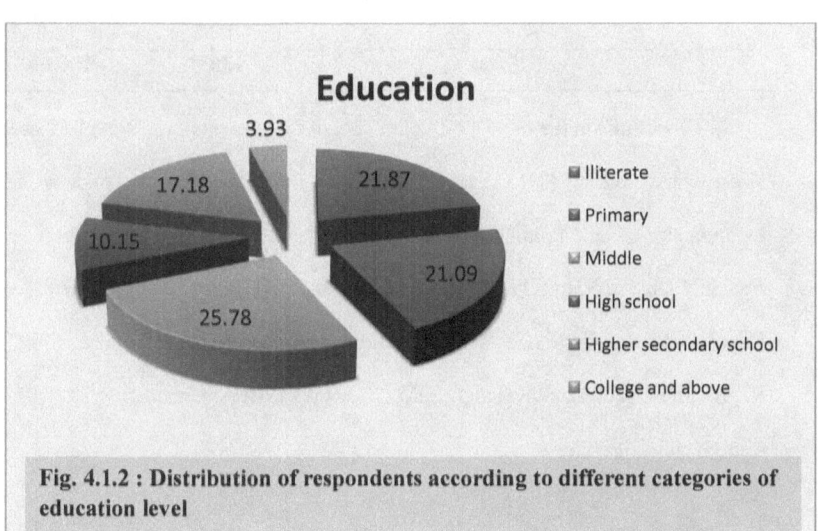

Fig. 4.1.2 : Distribution of respondents according to different categories of education level

The data presented in table 4.3 regarding size of family of respondents revealed that 63.28 per cent respondents belong to small size (i.e. up to 5 members), whereas, 32.81 per cent under medium size (6 to 10 members) and remaining 3.90 per cent under big size of family (more than 10 members). Study can be concluded that the majority of respondents belong to small size of family. Similar views have also been expressed by Singh *et al.* (2007), Dhruw (2008) and Lanjewar (2009).

4.1.4. Land holding:

Table 4.4: Distribution of sugarcane growers according to their land holding.

(n=128)

S.No.	Particulars	Frequency	Percent
1.	Marginal (up to 1 ha.)	29	22.66
2.	Small (1 to 2 ha.)	58	45.32
3.	Medium (2.1 to 4 ha.)	30	23.43
4.	Big (>4 ha.)	11	8.59
	Total	128	100.00

It seem from the table 4.4 that maximum respondents (45.31 %) had small size of land holding (1 to 2 ha.) followed by 23.43 per cent respondents had medium size of land holding (2.1 to 4 ha.), while 22.66 per cent respondents had marginal size of land holding (up to 1 ha.) and remaining 8.59 per cent respondents had big size of land holding (above 4 ha.). Nagraja (2002), Raghavendra (2004), Shivanand (2007), Itawdiya (2011), Chauhan *et al.* (2013) also given similar views.

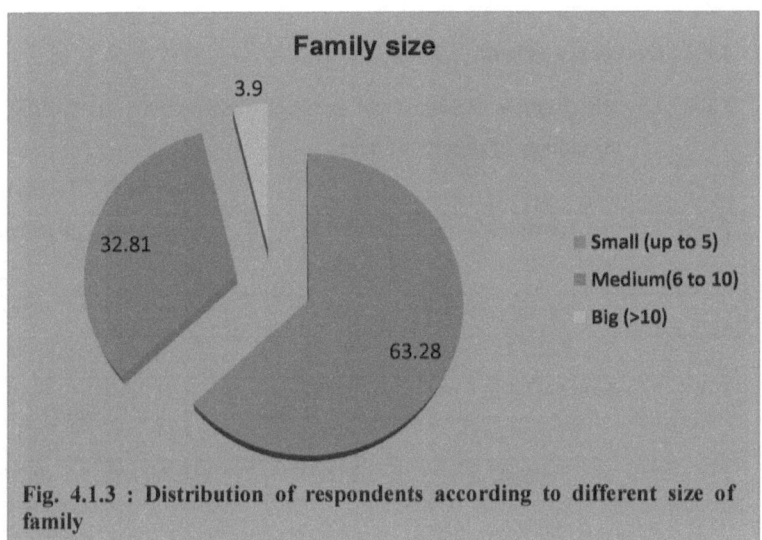

Fig. 4.1.3 : Distribution of respondents according to different size of family

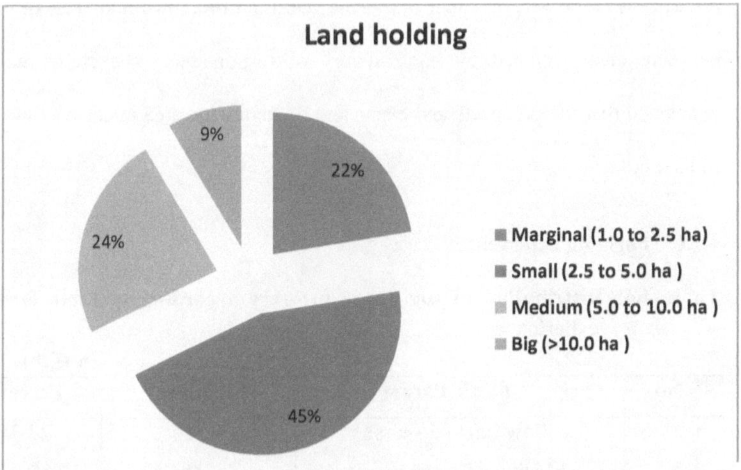

Fig. 4.1.4 : Distribution of respondents according to their land holding

4.1.5. Occupancy of land:

Table 4.5: Distribution of sugarcane growers according to their different occupancy of sugarcane area.

(Total land 272.25ha)

S.No.	Categories	Total land (ha.)	Sugarcane area (ha.)	Per cent
1.	Marginal (up to 1 ha.)	23.7	9.6	40.50
2.	Small (1.1 to 2 ha.)	95.41	40.79	42.75
3.	Medium (2.1 to 4 ha.)	85.81	36.91	43.01
4.	Big (>4 ha.)	67.96	23.47	34.53
	Total	272.25	110.77	40.24

It seem from the table 4.5 that out of the total occupied land, maximum 43.01 per cent sugarcane cultivated area occupied by the medium category of respondents followed by small (42.75%), marginal (40.50%) and rest of 34.53 per cent area occupied by big category of respondents. The study may be concluded that the prospects and expansion of sugarcane area under medium size of land holders.

4.1.6. Farming experience:

Table 4.6: Distribution of sugarcane growers according to their farming experience.

(n=128)

S.No.	Particulars	Frequency	Percent
1	Low (up to 5 years)	35	27.34
2	Medium (5 to 10 years)	41	32.04
3	High (>10 years)	52	40.62
	Total	128	100.00

The data presented in table (4.6) indicates that out of 128 respondents, 40.62 per cent respondents had high farming experience, followed by 32.03 per cent respondents had medium farming experience, whereas, remaining 27.34 per had cent had under low farming experience (up to 5 years).

Thus, it may be concluded that majority of tribal sugarcane growers (40.62%) have more than 10 years of farming experience. Above findings are also supported by Nagraja (2002), Maraddi (2006), Shivanand (2007) and Chauhan *et al.* (2013).

4.1.7. Annual income:

Table 4.7: Distribution of sugarcane growers according to their annual income. (n=128)

S.No.	Categories	Frequency	Percent
1.	Low (Up to Rs. 27,231)	3	2.34
2.	Medium (Rs. 27232 to 1,99324)	107	83.59
3.	High (>Rs. 1.99324.)	18	14.07
	Total	**128**	**100.00**

Annual income is an important indicator of the economic status of a family and can it plays a key role to overcome economic position. The distribution of the respondents according to their annual income presented in table 4.7 revealed that majority of respondents (83.59 %) belong to medium annual income (Rs. 27,232 to 1, 99,324), followed by 14.07 per cent of respondents under high annual income and only 2.34 per cent of respondents under low annual income (Up to Rs. 27,231). Similar views have also been expressed by Kanavi (2000), Nagaraja (2002) and Itawdiaya (2011).

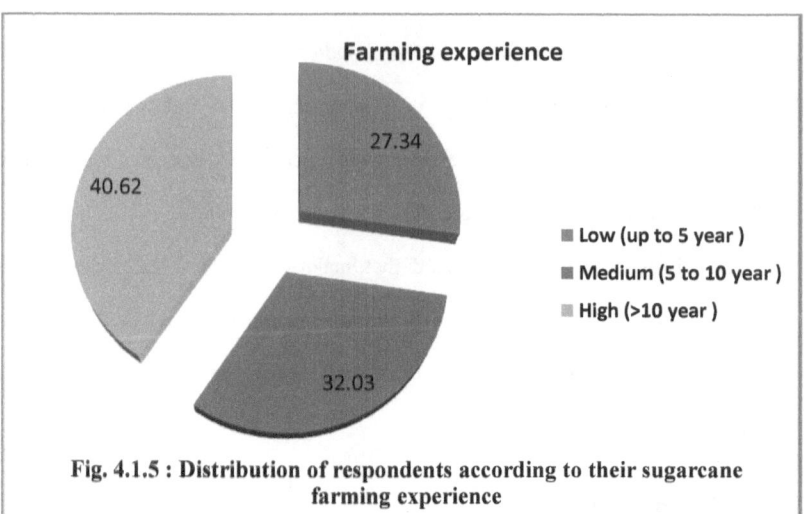

Fig. 4.1.5 : Distribution of respondents according to their sugarcane farming experience

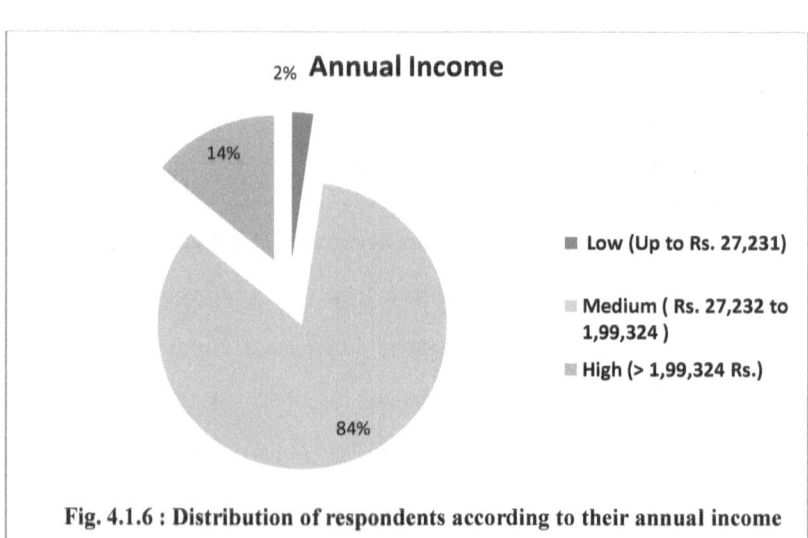

Fig. 4.1.6 : Distribution of respondents according to their annual income

4.1.8. Credit acquired:

The data presented in table 4.8 showed that 63.28 per cent respondents acquired credit, where as 36.71 per cent respondents not acquired the credit. Out of the credit acquiring respondents (total 81), the majority of the respondents (100.00%) had taken the short term credit and none of the respondent had acquired medium term and long term credit.

Table 4.8 Distribution of sugarcane growers according to their credit acquisition.

S. No.	Particulars	Frequency	Per cent
1.	**Credit acquired**		
	Not acquired	47	36.71
	Acquired	81	63.28
2.	**Duration of credit (n=81)**		
	Short term	81	100.00
	Medium term	00	00.00
	Long term	00	00.00
3.	**Source of credit (n=81)**		
	Co-operative society	00	00.00
	Nationalized bank	76	93.82
	Money lenders	00	00.00
	Friends/neighbours/ relatives/ others	05	6.18
4.	**Availability of credit (n=81)**		
	Easy	78	96.30
	Difficult	3	3.70

Majority of the respondents acquired short term credit to purchase seeds, fertilizers, implements etc and use to deposit their entire loan just after harvesting of crop to get further loan in next year again from same credit agency. This finding is in conformity with findings reported by Lanjewar (2009) and Chauhan (2013).

In case of sources of credit the maximum number of the respondents (93.82%) had acquired credit from the nationalized banks and 6.18 per cent respondents took credit from friends, neighbors and relatives etc.

The data presented in table 4.8 revealed that the respondents were aware about the sources of credit and facilities provided by nationalized banks. This finding is in conformity with findings reported by Dhruw (2008).

Regarding availability of credit, 96.30 per cent respondents acquired credit easily whereas, 3.70 per cent respondents faced some difficulty to acquired credit.

This may be concluded that majority of the respondents had acquired short term credit from nationalized banks whereas the major sources of credit and the credit facilities were available very easily and quickly. Limje (2000), Mukim (2004) and Lanjewar (2009) also reported similar findings.

4.1.9. Occupation:

Table 4.9 Distribution of respondents according to their occupation

(n=128)

S. No.	Occupation	Frequency	Per cent
1.	Agriculture	30	23.43
2.	Agriculture + Labour	94	73.43
3.	Agriculture + Service	4	3.13
4.	Agriculture + Animal husbandry	00	00
5.	Agriculture + business	00	00
6.	Agriculture + Other	00	00
	Total	128	100.00

The data presented in table 4.9 regarding the distribution of the respondents according to their occupation revealed that 73.43 per cent of the respondents were involved in agriculture work as well as labour followed by agriculture (23.43%) and agriculture + service (3.13%). It may be concluded that majority of the respondents were involved in agriculture work as well as labour. Findings find support with the work of Raghuvanshi (2003).

4.1.10. Social participation

Table 4.10: Distribution of respondents according to their Social participation.

(n=128)

S.No.	Particulars	Frequency	Per cent
1	No membership	95	74.21
2	Membership in one organization	21	16.42
3	Membership in more than one organization	5	3.90
4	Executive/ office bearer	7	5.47
	Total	128	100.00

Social participation provides an idea about the respondent's participation in social activities. Regarding social participation, maximum number of

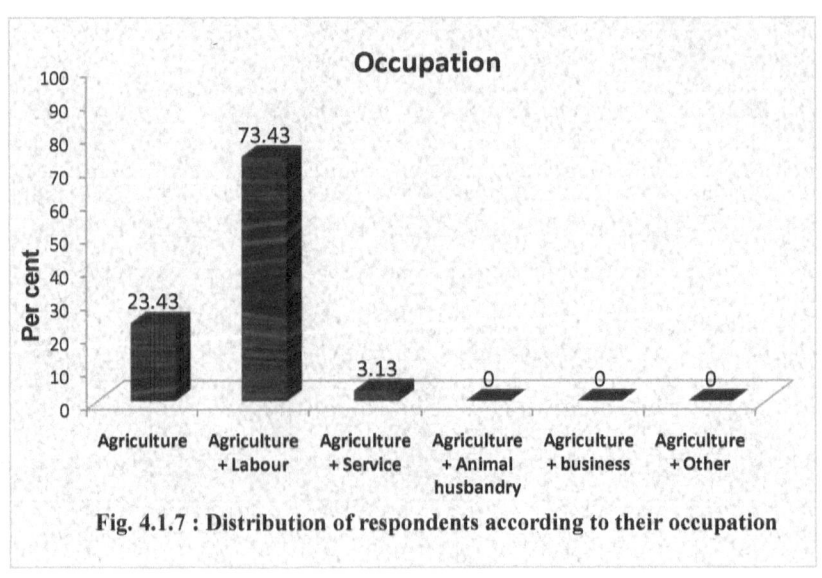

Fig. 4.1.7 : Distribution of respondents according to their occupation

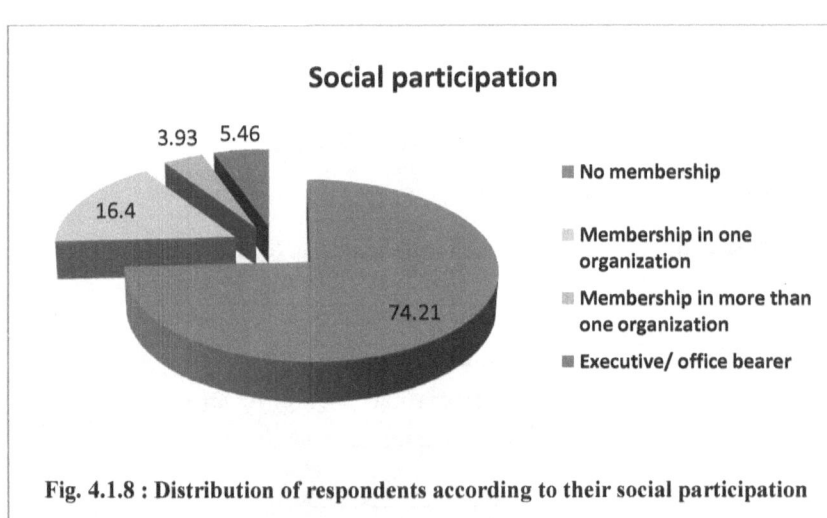

Fig. 4.1.8 : Distribution of respondents according to their social participation

respondents (74.21%) had no membership in any organization followed by 16.45 per cent of respondents who had membership in one organization. Only 5.47 per cent respondents who belonged to executive/ office bearer category while only 3.90 per cent had membership in more than one organization.

It can be concluded that majority of the respondents had no membership in any organization. This finding is in conformity with findings reported by Sharma et al. (2000), Choudhary (2003), Dhruw (2008) and Varma (2009).

4.1.11. Scientific-orientation:

The results in the Table 4.11 showed that majority of the respondents (67.98%) had medium level of Scientific–orientation, followed by 18.75 per cent had high level of scientific–orientation, while, 18.75 per cent had low level of scientific-orientation regarding sugarcane production technology.

Table4.11: Distribution of respondents according to their scientific orientation.   (n= 128)

S.No	Categories	Frequency	Per cent
1.	Low (26.40)	17	13.28
2.	Medium (26.41 to 29.15)	87	67.98
3.	High(>29.15)	24	18.75
	Total	128	100.00

$\bar{X} = 27.78$ S.D= 1.38

It can be concluded that majority of the respondents came under the medium level of scientific–orientation category. This finding is in similar to the findings reported by Palaniswamy (2001), Nagaraja (2002), Maraddi (2006) and Shivanand (2007).

4.1.12. Contact with extension agency:

The result of table 4.12 indicate that (58.59%) respondents had medium level of overall contact with extension agency, followed by 23.45 per cent respondents had low level of contact with extension agency and only 17.96 per cent respondents had high level of contact with extension agency.

Table 4.12: Distribution of respondents according to overall contact with extension agency. (n= 128)

S.No.	Categories	Frequency	Per cent
1.	Low (1-9.33 sources)	30	23.45
2.	Medium(9.33-12.63 sources)	79	61.71
3.	High(> 12.63 sources)	19	14.84
	Total	128	100.00

$\bar{X}$= 7.42 　　　　　　　　　　　　S.D= 2.72

It can be concluded that majority of the respondents came under the medium level of overall contact with extension agency. Findings find support with the findings of Panday (1996).

Table4.13: Distribution of respondents according to their contact with individual extension agency.

S.No	Extension Agency	No contact	Yearly (2-3 times)	Monthly	Weekly
1.	Rural Agriculture Extension Officer	0 (0.000)	0 (0.00)	4 (3.13)	124 (96.88)
2.	Agruculture Development Officer	89 (69.53)	19 (14.84)	20 (15.63)	0 (0.00)
3.	Sugarcane Development Officer	49 (38.28)	37 (28.91)	42 (32.81)	0 (0.00)
4.	Subject Matter Specialist (KVK)	100 (78.13)	23 (17.97)	5 (3.90)	0 0.00
5.	Agriculture Scientist	91 (71.10)	29 (22.66)	8 (6.25)	0 (0.00)
6.	NGO's	128 (100.00)	0 (0.00)	0 (0.00)	0 (0.00)

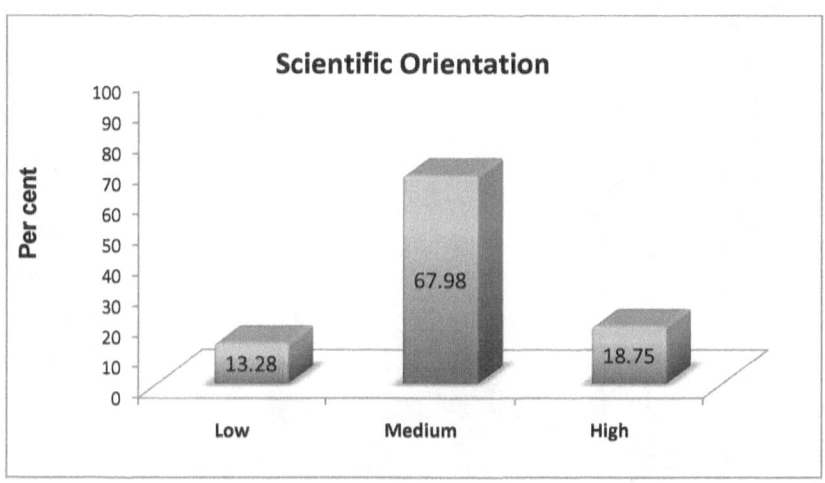

Fig. 4.1.9 : Distribution of respondents according to their scientific orientation

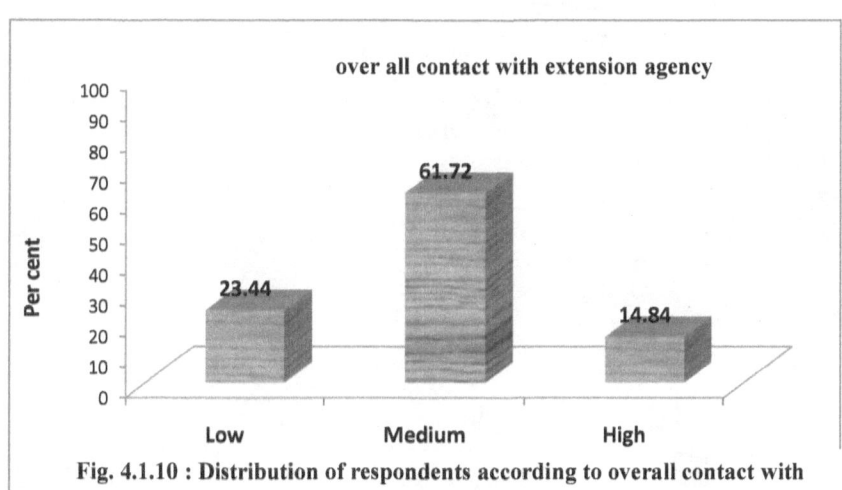

Fig. 4.1.10 : Distribution of respondents according to overall contact with extension agency

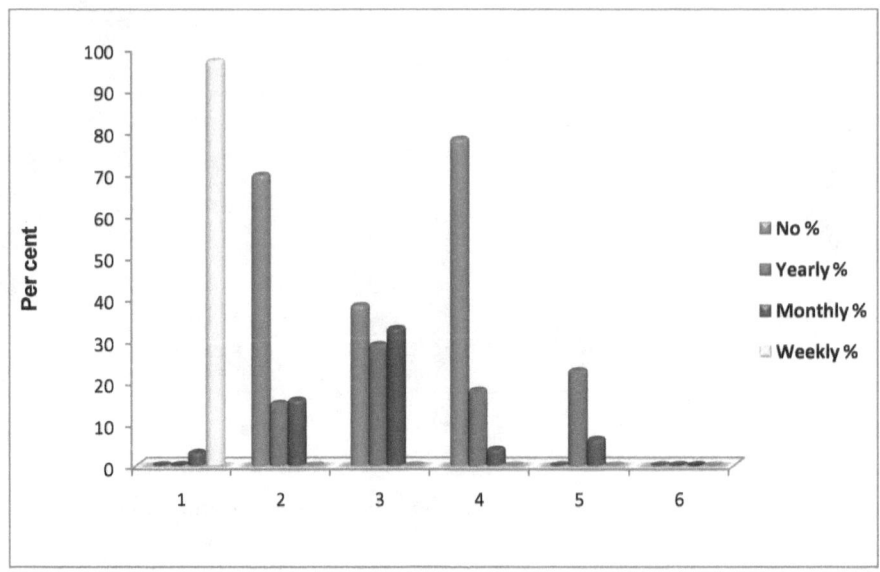

Fig. 4.1.11 : Distribution of respondents according to their contact with individual extension agency

1. Rural Agriculture Extension Officer
2. Agriculture Development Officer
3. Sugarcane Development Officer
4. Subject Matter Specialist (KVK)
5. Agriculture Scientist (Agriculture college's)
6. NGO's

The data presented in table 4.13 regarding contact of respondents with extension agency revealed that majority of the respondents (100.00%) never consulted NGOs to obtain information about the recommended sugarcane production technology, whereas 78.13 per cent respondents never contacted subject matter specialist (KVK), 71.10 per cent respondents never contacted Agriculture Scientists, 69.53 per cent and 38.28 per cent of the respondents never contacted extension agency *viz.* Sugarcane development officer (sugar factory), and ADOs respectively for obtaining information about recommended sugarcane production technology.

Majority of the respondents (28.91%) consulted sugarcane development officer (sugar factory) ones in a year for getting information about recommended sugarcane production technology, whereas, 22.66 per cent consulted Agriculture scientists and 17.97 per cent contacted subject matter specialist (KVK). only 14.84 per cent of respondents contacted Agriculture Development officer (ADOs) and never contacted with (RAEOs).

Few respondents (32.81%) consulted Sugarcane Development officer once in a month to know about the new production technology of sugarcane and related information, while 15.63 per cent of the respondents got valuable suggestion from ADOs once in a month, whereas, (6.25%) respondents consulted Agriculture scientists in a month to get information about recommended sugarcane production technology. While 3.90 per cent of respondents contacted subject matter specialist (KVK) and only (3.13%) of the respondent contacted RAEOs once in a month to get aware about recommended sugarcane production technology.

The data presented in table 4.13 also revealed that majority of respondents (96.88%) had contacted RAEOs regular (weekly) for getting new information related to sugarcane production technology.

It may be concluded that RAEOs (Agriculture Department) and Sugarcane Development officer (Sugar Factory) were the most frequently visiting extension personnel in the villages from which the respondents obtained latest information regarding recommended sugarcane production technology. Findings find support with the work of Mazher *et al.* (2003).

4.1.13. Use of information sources

Table 4.14: Distribution of respondent according to overall use of information source

(n= 128)

S.No.	Categories	Frequency	Per cent
1.	Low	19	14.84
2.	Medium	91	71.09
3.	High	18	14.06
	Total	128	100.00

$\bar{X} = 17.79$ SD = 2.36

The findings presented in table 4.14 revealed that majority (71.09%) of the respondents utilized medium level of information sources, followed by 14.84 per cent had utilized low level of information sources, while 14.06 per cent had utilized high level of sources of information.

Table 4.15: Distribution of respondents according to their use of information sources

S.No.	Sources of information	Frequency	Percentage*	Ranks
1.	Progressive farmer	113	88.28	I
2.	Neighbor	105	82.03	III
3.	Friends	27	21.09	VII
4.	Local leaders/ Sarpanch	1	0.78	XII
5.	Relatives	1	0.78	XII
6.	Co-operative society	4	3.12	X
7.	Agriculture magazine	36	28.12	VI
8.	Radio	107	83.59	II
9.	Television	86	67.18	IV
10.	Farmer fair	8	6.25	IX
11.	Training	40	31.25	V
12.	Farm visit	15	11.71	VIII
13.	Leaflet/Pamphlet	2	1.56	XI

*Data are based on multiple responses

About 88.28 per cent of the respondents had received the information from progressive farmers followed by 83.59 per cent of the respondents had obtained information from radio, 82.03 per cent from television, 31.25 per cent obtained information through training, 28.12 per cent from agriculture magazine, (21.09%) from friends, 11.71 percent of the respondents had obtained information from farm visit, 6.25 per cent from farmer fair, 1.56 per cent of the respondents had obtained information through leaflet/pamphlet, while only 0.78 per cent of the respondent obtained information from relatives and local leaders/sarpanch. Painkra (2000), Rajni (2006) and Suryawanshi (2009) have also reported similar findings.

Table 4.15: Distribution of respondents according to their use of information sources

S.No.	Sources of information	Frequency	Percentage*	Ranks
1.	Progressive farmer	113	88.28	I
2.	Neighbor	105	82.03	III
3.	Friends	27	21.09	VII
4.	Local leaders/ Sarpanch	1	0.78	XII
5.	Relatives	1	0.78	XII
6.	Co-operative society	4	3.12	X
7.	Agriculture magazine	36	28.12	VI
8.	Radio	107	83.59	II
9.	Television	86	67.18	IV
10.	Farmer fair	8	6.25	IX
11.	Training	40	31.25	V
12.	Farm visit	15	11.71	VIII
13.	Leaflet/Pamphlet	2	1.56	XI

*Data are based on multiple responses

About 88.28 per cent of the respondents had received the information from progressive farmers followed by 83.59 per cent of the respondents had obtained information from radio, 82.03 per cent from television, 31.25 per cent obtained information through training, 28.12 per cent from agriculture magazine, (21.09%) from friends, 11.71 percent of the respondents had obtained information from farm visit, 6.25 per cent from farmer fair, 1.56 per cent of the respondents had obtained information through leaflet/pamphlet, while only 0.78 per cent of the respondent obtained information from relatives and local leaders/sarpanch. Painkra (2000), Rajni (2006) and Suryawanshi (2009) have also reported similar findings.

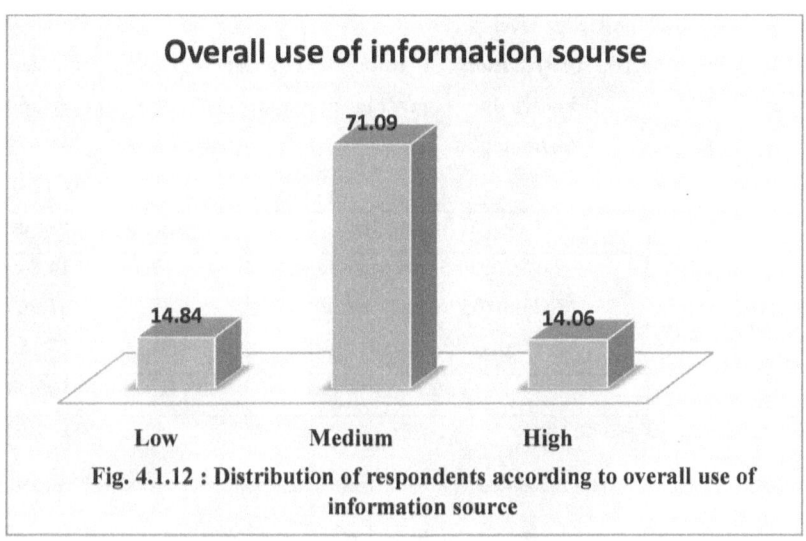

Fig. 4.1.12 : Distribution of respondents according to overall use of information source

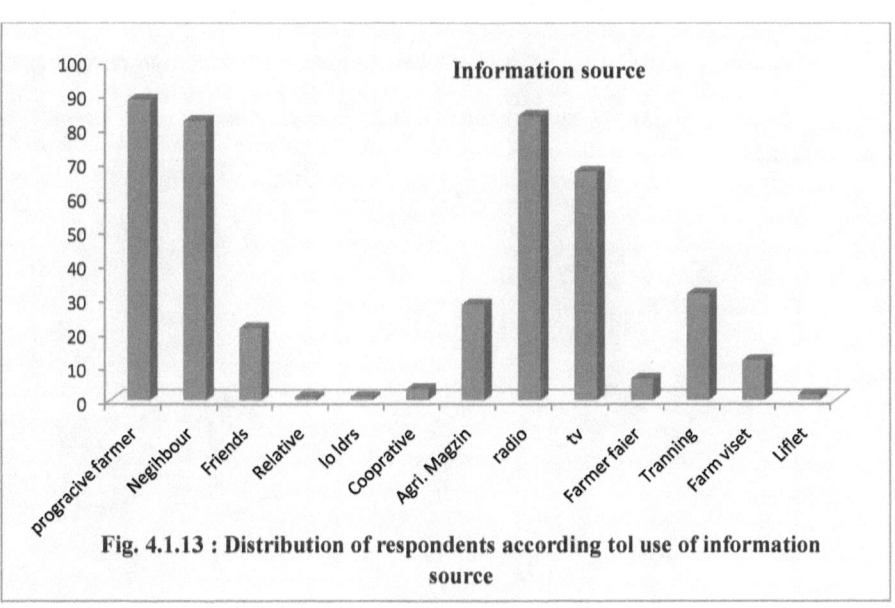

Fig. 4.1.13 : Distribution of respondents according tol use of information source

4.1.14. Knowledge level

Table 4.16: Distribution of respondent according to overall level of knowledge regarding recommended sugarcane production technology

(n= 128)

S.No.	Level of knowledge	Frequency	Per cent
1.	Low (up to 34.98 score)	19	14.84
2.	Medium (35 to 40.37 score)	87	67.96
3.	High (>40.38 score)	22	17.18
	Total	128	100.00

$\bar{X} = 37.67$ SD =2.63

The data presented in table 4.16 revealed that out of total 128 respondents, 67.96 per cent respondents indicated medium level of knowledge, followed by (17.18%) had high level of knowledge, while only 14.84 per cent found having low level of knowledge for the sugarcane production technology. Findings find support with the work of Naik (2005), Maraddi (2006), Shivanand (2007), Itawdiya (2011) and Chauhan *et al.* (2013).

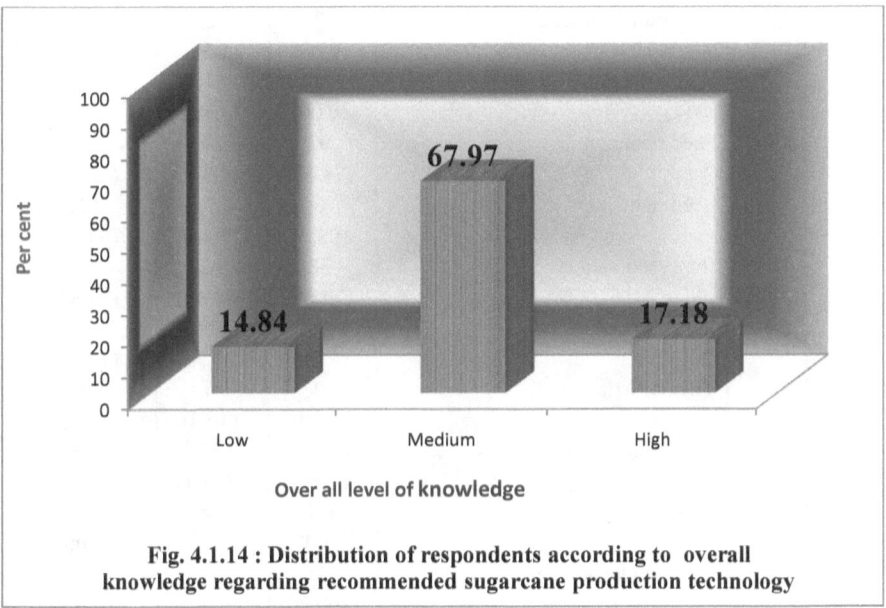

Fig. 4.1.14 : Distribution of respondents according to overall knowledge regarding recommended sugarcane production technology

Table 4.17: Distribution of respondent according to their practices- wise level of knowledge regarding recommended sugarcane production technology. (n=128)

S.No.	Sugarcane cultivation practices	Level of knowledge		
		Low f(%)	Medium f(%)	High f(%)
1.	Selection of land	3 (2.34)	50 (39.06)	75 (58.59)
2.	Preparation of land	0 (0.00)	2 (1.56)	126 (98.46)
3.	Seed selection	0 (0.00)	4 (3.12)	124 (96.88)
4.	Seed treatment	97 (75.78)	3 (2.34)	28 (21.88)
5.	Seed rate	0 (0.00)	7 (5.46)	121 (94.53)
6.	Improved variety	38 (29.69)	62 (48.43)	28 (21.87)
7.	Fertilizer use	1 (0.78)	99 (77.34)	28 (21.87)
8.	Time of irrigation	0 (0.00)	6 (4.68)	122 (95.32)
9.	Weed management	2 (1.56)	53 (41.41)	73 (57.03)
10.	Insect pest management	110 (85.93)	18 (14.06)	0 (0.00)
11.	Disease management	128 (100.00)	0 (0.00)	0 (0.00)
12.	Earthing up	0 (0.00)	16 (12.5)	112 (87.5)
13.	Tying	109 (85.16)	1 (0.78)	18 (14.06)
14.	Harvesting	0 (0.00)	14 (10.93)	114 (89.06)
15.	Marketing	0 (0.00)	0 (0.00)	128 (100.00)
16.	Ratoon management	2 (1.56)	6 (4.68)	120 (93.75)

F= Frequency   %= Percent

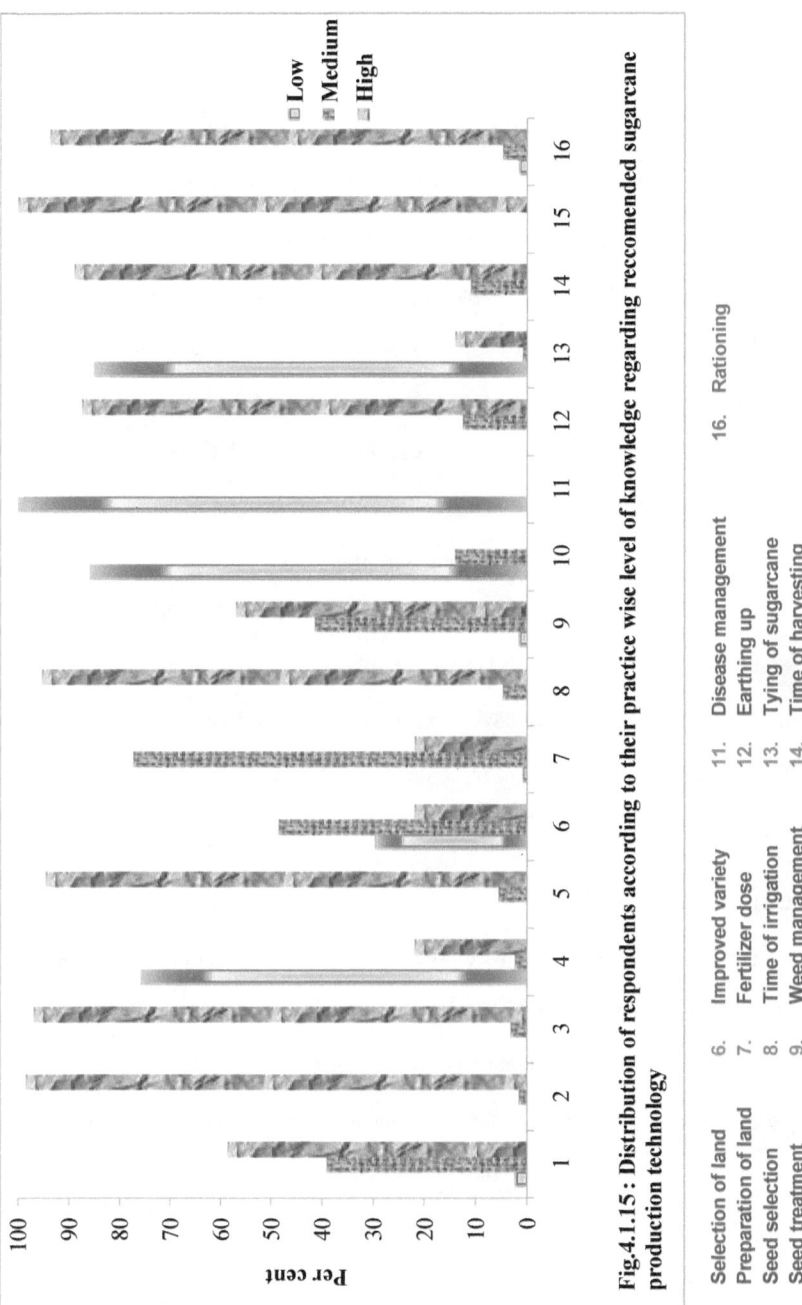

Fig.4.1.15 : Distribution of respondents according to their practice wise level of knowledge regarding reccomended sugarcane production technology

1. Selection of land
2. Preparation of land
3. Seed selection
4. Seed treatment
5. Seed rate
6. Improved variety
7. Fertilizer dose
8. Time of irrigation
9. Weed management
10. Insect management
11. Disease management
12. Earthing up
13. Tying of sugarcane
14. Time of harvesting
15. Marketing
16. Rationing

The data presented in the table 4.17 revealed that majority of the respondents had low level of knowledge regarding selected 16 practices of sugarcane production technology *i.e.* disease management (100.00%), insect-pest management (85.93%), tying of sugarcane (85.16%), seed treatment (75.78%), improved variety (29.69). Whereas, the majority of the respondents were having medium level of knowledge regarding sugarcane production technology *i.e.* fertilizer use (77.34%), improved variety (62.43%), weed management (41.41%), selection of land (39.06%), insect-pest management (14.06), earthing up (12.05%), harvesting time (10.93%), seed rate (5.46). While respondents high level of knowledge group selected practices is like marketing facility (100.00%), preparation of land (98.46%), seed selection (96.88%), time of irrigation (95.32%), seed rate (94.53%), ratoon management (93.75%), harvesting (89.06%), earthing up (87.05%), selection of land (58.59%), weed management (57.03%), seed treatment (21.88%), improved variety and fertilizer use (21.87%), tying of sugarcane (14.06%), of respondent were having high level of knowledge respectively

4.2 Dependent variable:

4.2.1: Extent of adoption of recommended sugarcane production technology by tribal sugarcane growers:

Over all extent of adoption is clearly indicated from the data presented in Table 4.17, that out of total respondents, majority (60.94%) had medium level of adoption of recommended sugarcane production technology. Whereas, 19.53 and 19.53 per cent of them had high and low level of adoption, respectively.

Table 4.18: Distribution of respondents according to over all extent of adoption regarding recommended sugarcane production technology: (n= 128)

S.No.	Level of adoption	Frequency	Per cent
1.	Low (up to 34.04 score)	25	19.53
2.	Medium (34.05 to 38.73 score)	78	60.94
3.	High (>40.38 score)	25	19.53
	Total	**128**	**100.00**

$\bar{X}$= 36.57   SD=2.16

It is concluded from the table 4.18 that majority (60.94%) of the respondents showed medium level of adoption regarding recommended sugarcane production technology. Whereas, 19.53 per cent of the respondents showed high level of adoption. Medium to high adoption may be due to the facts that the respondents were educated, possessed large land holdings, belonged to higher income group had better utilization of information sources such as friends, relatives, neighbors, radio, progressive farmers etc. and better orientation towards scientific technologies. The findings find support with the work of Kanavi (2000), Nagaraja (2002), Pawar (2005), Naik (2005), Mishra (2006), Itawdiya (2011) and Chouhan (2013).

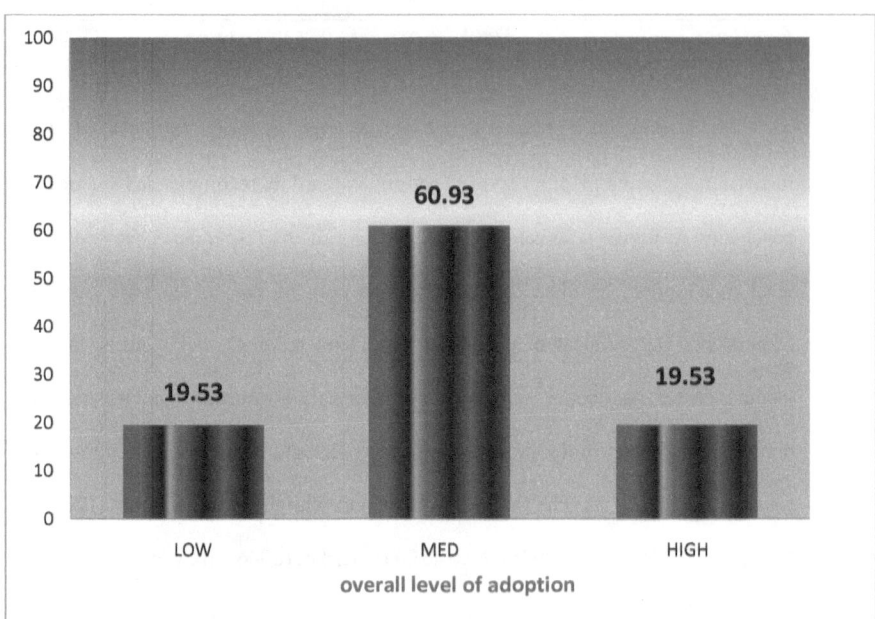

Fig. 4.1.16 : Distribution of respondents according to overall adoption regarding recommended sugarcane production technology

Table 4.19: Adoption level of sugarcane growers according to their different improved sugarcane cultivation practices:

(n=128)

S.No.	Sugarcane cultivation practices	Level of Adoption		
		Low f (%)	Medium f (%)	High f (%)
1.	Selection of land	6 (4.68)	92 (71.87)	31 (24.21)
2.	Preparation of land	0 (0.00)	2 (1.56)	126 (98.43)
3.	Seed selection	1 (0.78)	1 (0.78)	126 (98.43)
4.	Seed treatment	109 (85.15)	3 (2.34)	16 (12.5)
5.	Seed rate	0 (0.00)	6 (4.68)	122 (95.31)
6.	Improved variety	30 (23.43)	69 (53.90)	29 (22.65)
7.	Fertilizer use	0 (0.00)	112 (87.5)	16 (12.5)
8.	Time of irrigation	0 (0.00)	16 (12.5)	112 (87.5)
9.	Weed management	3 (2.34)	57 (44.53)	68 (53.12)
10.	Insect management	116 (90.63)	12 (9.37)	0 (0.00)
11.	Disease management	128 (100.00)	0 (0.00)	0 (0.00)
12.	Earthing up	0 (0.00)	16 (12.5)	112 (87.5)
13.	Tying	114 (89.06)	4 (3.12)	10 (7.81)
14.	Harvesting	0 (0.00)	45 (35.15)	83 (64.84)
15.	Marketing	0 (0.00)	0 (0.00)	128 (100.00)
16.	Ratoon management	1 (0.78)	13 (10.16)	114 (89.06)

F= Frequency   %= Percent

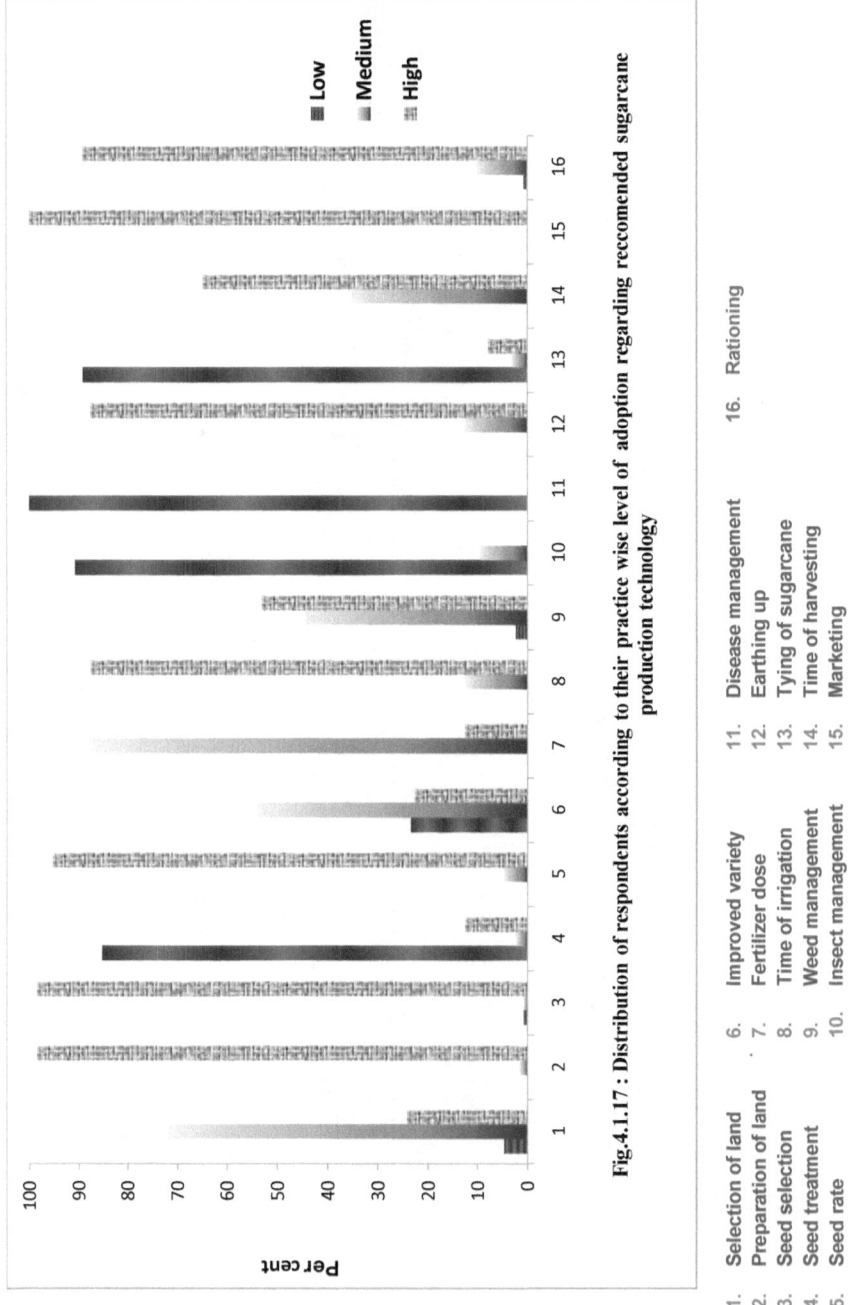

Fig.4.1.17 : Distribution of respondents according to their practice wise level of adoption regarding reccomended sugarcane production technology

1. Selection of land
2. Preparation of land
3. Seed selection
4. Seed treatment
5. Seed rate
6. Improved variety
7. Fertilizer dose
8. Time of irrigation
9. Weed management
10. Insect management
11. Disease management
12. Earthing up
13. Tying of sugarcane
14. Time of harvesting
15. Marketing
16. Rationing

Adoption level of improved sugarcane cultivation practices was studied in term of sixteen recommended practices. The data collected, analyzed and presented in table 4.19

Regarding field selection, 71.87 per cent respondents had under medium level of adoption followed by 24.21 per cent had high level of land selection and 4.68 per cent had low level of adoption in the practices of land selection.

The majority of the respondents (98.43%) had adopted the recommended technology of land preparation, while 1.56 per cent had medium level adoption of field preparation.

Regarding seed selection, 98.43 per cent respondents had high level of adoption while 1.56 per cent had medium and low adoption category.

The majority of growers (85.15%) had not adoption the seed treatment, while 2.34 per cent had medium and only 12.5 per cent showed high level of adoption for seed treatment practices.

Regarding recommended seed rate, majority of (95.31) respondents had high level adoption for recommendation seed rate, while 4.68 per cent respondents showed only medium level adoption.

The majority of sugarcane growers (53.90%) had medium level adoption for improved recommended varieties, while 22.65 per cent respondents had high level adoption and 23.43 per cent not adopted improved varieties of sugarcane.

The most of the respondents (87.05%) showed medium level adoption in fertilizer application, while 12.05 per cent had high level of adoption in fertilizer application.

The majority of respondent (87.5%) indicate high level adoption for time of irrigation management while 12.05per cent respondents had medium level adoption for time of irrigation management.

The most of the respondents (53.12%) showed high level adoption for weed management, and 44.53 per cent had medium level of adoption of weed management and 2.34 per cent had low level adoption of weed management.

Regarding insect pest management, higher percentage of respondents (90.63%) under low level of adoption, while 12.37 per cent showed medium level of adoption of insect pest management.

The cant percent respondents (100.00%) had low level of adoption of disease management.

Regarding earthing up of sugarcane, most of sugarcane growers (87.5%) showed high level of adoption, while 12.05 per cent had medium level of adoption.

The majority of sugarcane growers, 89.06 per cent had low level of adoption in tying of sugarcane while 7.81 per cent had high level adoption and (3.12%) had medium level adoption category of sugarcane tying.

Regarding harvesting time, higher percentage of respondents (64.84%) showed high level of adoption, while 35.15% respondents had medium level of adoption category of sugarcane harvesting in time.

The cant percent of respondents (100.00%) indicated higher level of marketing of sugarcane at sugar factory.

The majority of respondents (89.06%) showed high level of adoption in ratoon management, while 10.16 per cent respondents had medium level of adoption and remaining 0.78 per cent had low level of ratoon management.

4.3 Correlation analysis of independent variables with adoption of recommended sugarcane production technology

Table 4.20: Correlation analysis of independent variables with extent of adoption of sugarcane production technology

S.No.	Independent variable	Correlation coefficient (r)
1	Age	-0.201 NS
2	Education	0.173314 NS
3	Family size	0.027845 NS
4	Occupation	-0.09309 NS
5	Land holding	0.165167 NS
6	Experience	0.282**
7	Annual income	0.307*
8	Credit acquisition	0.525**
9	Scientific orientation	0.202529 NS
10	Social participation	-0.05676 NS
11	Information source	0.167223 NS
12	Contact with extension agency	0.281*
13	Level of knowledge	0.628**

** Significant at 0.01 level of probability; NS= Non- significant
* Significant at 0.05 level of probability;

Correlation coefficient between the selected characteristics of the respondents with adoption of recommended sugarcane production technology by tribal sugarcane growers have been worked out and presented in Table 4.20. It may be concluded from the result that out of all selected profiles characteristics, correlation coefficient between experiences, level of knowledge and Credit acquisition had found positive and highly significant at 0.01 level of probability. However, the variable annual income and contact with extension agencies were found positive and significant relation with extent of adoption at 0.05 percent level of significance. The variables, age, occupation and cosmopolitans were found to be negatively correlation with adoption, while education, family size, land holding, scientific orientation and source of information showed no significant relationship with extent of adoption of recommended sugarcane production technology by tribal sugarcane growers.

4.4 Multiple regression analysis of independent variables with adoption of recommended sugarcane production technology

Multiple regression analysis is employed to find out relative contribution of independent variables towards the dependent variables the result of regression analysis are presented in table 4.21.

Table 4.21: Multiple regression analysis of independent variables with the adoption of recommended sugarcane production technology:

S.No.	Independent variable	Regression coefficient (b)	't' value
1	Age	-0.495 NS	-1.795
2	Education	-0.081 NS	-0.72
3	Family size	0.186 NS	0.657
4	Occupation	0.168 NS	0.448
5	Land holding	-0.223 NS	-0.955
6	Experience	0.203*	0.865
7	Annual income	0.402 NS	1.404
8	Credit acquisition	0.222**	4.543
9	Scientific orientation	0.117 NS	1.021
10	Social participation	-0.603 NS	-2.285
11	Information source	-0.021 NS	-0.464
12	Contact with extension agency	0.155*	0.806
13	Level of knowledge	0.477**	6.84

** Significant at 0.01 level of probability; $R^2 = 0.463$
* Significant at 0.05 level of probability; F value of R = 8.20

NS= Non- significant

The result of multiple regression analysis revealed that out of 13 independent variables, level of knowledge contributed highly significant and Credit acquisition at 0.01 percent level of probability, whereas, experience, contact with extension agency were found significant to the adoption at 0.05 percent level of significance as evident from significant 't' value of these variables *viz.* experience, contact with extension agency and level of knowledge would cause 0.203, 0.155 and 0.477 unit changes, respectively in adoption of recommended sugarcane production technology by the tribal sugarcane growers.

From Table 4.21, it could be further seen that all the 12 independent variables jointly explained the variation to the extent 46.3 percent in adoption of recommended sugarcane production technology, which was found statistically significant at 0.01 percent and 0.05 percent level of significant as could be seen form calculated 'F' value of 8.20 it is therefore suggested that if the rate of adoption of recommended sugarcane production technology by tribal sugarcane growers wanted to be hiked there are need to be accelerated their frequency of contact with extension agency and level of knowledge through proper guidance and training. This will be helpful for increasing the adoption of recommended sugarcane production technology.

4.5 Constraints faced by the sugarcane growers in adoption of recommended sugarcane production technology

Multiple responses were taken to ascertain the constraints faced by the sugarcane growers in adoption of recommended sugarcane production technology. Various problems are presented in Table 4.22 which indicates that under the personal constraints, maximum number of respondents (56.00%) reported scattered land, followed by lack of education (15.62%), other occupation (6.23%) and over age of farmer (3.90%) as the major constraints.

Table 4.22: Constraints faced by the sugarcane growers in adoption of recommended sugarcane production technology

S. No.	Constraints	Frequency*	Per cent	Rank
A	Personal Constraints			
1.	Low level of education	22	15.62	II
2.	Over age of respondents	5	3.90	IV
3.	Other occupations	18	6.25	III
4.	Scattering of existing land	72	56.00	I
B	Socio economic Constraints			
1.	Small size of land	58	45.31	IV
2.	Low income	38	29.68	VI
3.	Credit facilities is not available at proper time	27	21.09	VII
4.	Inadequate credit	47	36.71	V
5.	Lack of risk bearing ability	73	57.03	III
6.	Lack of agriculture labors	91	71.09	II
7.	Damage of sugarcane crop by animals (elephant, rat and wild bear)	105	82.03	I
C	Socio- psychological Constraints			
1	Lack of social participation	56	43.75	II
2	Lack of motivation towards new technology	70	54.68	I
D	Communicational and information Constraints			
1.	Non availability of information in proper time	75	58.60	II
2.	Inadequate information	84	65.62	I

3.	Non-availability of information sources in proper time	65	50.78	III
4.	Less contact with extension officer	56	43.75	IV
E	**Technical Constraints**			
1.	Non-availability of agricultural inputs (Seed, Fertilizer and Insecticides etc.)	80	62.05	V
2.	Lack of knowledge about the improved technology	72	56.25	VI
3.	Non availability of electricity at proper time	92	71.87	III
4.	Lack of knowledge about of sugarcane tying	110	85.93	II
5.	Lack of knowledge about insects identification and management	118	92.18	I
F	**Transport Constraints**			
1.	Non availability, insufficient quantity of truck and tractor at proper time	80	62.05	I
2.	High cost of transporting charge	71	55.46	II

* Frequency based on multiple responses

It has been observed from the Table 4.22 that major constraints encountered by them under socio-economic constraints, damage of sugarcane crop by the wild animal (82.03%), lack of agriculture labour (71.09%), lack of risk bearing ability (57.03%), small size of land (45.31%), inadequate credit (36.71%), low income (29.68%), credit facilities not available at proper time (21.09%).

In case of social psychological constraints, the respondents reported to lack of motivation towards new technology (54.68%), and lack of social participation (43.75%) as the major constraints among tribal sugarcane growers.

In case of communicational constraints, the maximum respondents reported less contact with extension officer (43.75%), Non-availability of information sources in proper time (50.78%), Non availability of information in proper time (58.60%) and Inadequate information (65.62%) as major important constraints.

In case of technical constraints, maximum number of the respondents (92.18%) reported that lack of knowledge about insects identification and treatment, followed by lack of knowledge about of sugarcane tying (85.93%), non availability of electricity at proper time (71.87%), lack of knowledge about seed treatment methods (69.53%), non availability of agriculture inputs (seeds, fertilizers etc.) 62.05 per cent and lack of knowledge about the adoption of improve technology (56.25%), as the major constraints among sugarcane growers. The findings find support from Channal (1995), Karthikeyan *et al.* (1996), Kanavi (2002), Nagaraja (2002), Shivanand (2007) and Chauhan *et al.* (2013).

4.6 Suggestions offered by the tribal sugarcane growers to minimize the constraints faced by them

From the table 4.23 on suggestion to overcome by tribal sugarcane growers for adoption of improved sugarcane cultivation practices were dissemination of information or technological knowledge should be given by RAEOs and sugarcane development officer 92.18 per cent. This is essential

Table 4.23: Suggestions of tribal sugarcane growers for minimizing the constraints faced by them during the adoption of recommended sugarcane production technology

S.No.	Suggestion	Frequency	Percent	Rank
1.	Rate of seed and fertilizer should be less	104	81.25	IV
2.	Electricity should be made available	100	78.12	V
3.	Demonstration should be conducted on farmers field by agriculture department	86	67.87	VIIII
4.	Sugarcane training should be organized	90	70.31	VII
5.	Technological knowledge should be given by REAO's and sugarcane development officer.	118	92.18	I
6.	The crop loan and subsidy should be provided in proper time	107	83.59	III
7.	Field visit should be taken by ADO's twice in a month	96	75.00	VI
8.	Provide improved varieties by the government agencies	112	87.05	II
9.	Provide field fencing subsidy by the government	89	69.53	VIII

*Frequency based on multiple respondents

Requirement for increasing the sugarcane cultivation. Other suggestion in order of rank were, provide improved varieties by the government agencies, the crop lone and subsidy should be provided in proper time (83.59%), cost of seed & fertilizer should be less (81.25%), electricity should be made available (78.12%), field visit should be taken by ADOs at least twice in a month (75.00%), sugarcane training should be organized (70.31 per cent), provide subsidy fencing of farm or field by the government (69.53%) and demonstration should be conducted on farmers field by department of agriculture (67.87%).

Form the table 4.23 on suggestion to overcome by tribal sugarcane growers for adoption of improved sugarcane cultivation practices were, dissemination of information of technological knowledge should be given by REAOs and sugarcane development officer 92.18 per cent. This is essential requirement for increasing the sugarcane cultivation. Other suggestion in order of rank were, provide improved varieties by the government agencies, the crop lone and subsidy should be provided in proper time (83.59%), cost of seed, fertilizer & should be less (81.25%), electricity should be made available (78.12%), field visit should be taken by ADOs twice in a month (75.00%), sugarcane training should be organized 70.31 per cent, provide field fencing subsidy by the government(69.53%) and demonstration should be conducted on farmers field by Department of Agriculture (67.87%).

Researcher obtaining the information from the respondents

Researcher obtaining the information from the respondents

Sugarcane processing factory

Summary, Conclusion & Suggestions for Future Research Work

CHAPTER-V

SUMMARY, CONCLUSION AND SUGGESTIONS FOR FUTURE RESEARCH WORK

The main purpose of this chapter is to summaries the results to state the conclusions on the basis of the foregoing analysis and to indicate some of their implications for actions.

Sugarcane is one of the oldest crops being cultivated in India. India occupies the first rank in production of sugarcane in the world. However, it ranks only 10th in world productivity .though enough viable and adoptive technologies of its cultivation have been developed, there enlists a wide adoption gap among the farmers. In consequence to this the production of sugarcane in the country 342.20 million tones in 2011 – 12. In Chhattisgarh sugarcane production 45.42 thousand tones during 2011-12.

The development of the agriculture is basically the application of science and technology by making the best use of available resources. The new agricultural technology is the result of solid foundation and meticulous planning in agriculture research. Paramount problem is the uneven distribution of benefits of modern technology. Process of transfer of modern technology has been allowed and not uniform in various parts of the country, hardly 20 per cent of the available modern technology adopted so far. The production and productivity of sugarcane increased by adoption of new technology followed by modernization sugarcane as well as promote farmer to grow more the crop.

The sugarcane productivity has shown an increasing trend over the year. The magnitude has been quite trivial, wide gap exists between potential and the realized productivity. The gap between potential yield and realized yield is due to environmental factors and socio-economic factors, induced cropping system, available varieties, fertilizer application, pest and disease management, marketing, post harvest problems and also communicational factors.

Although crop improvement projects have served by Indian agriculture admirable by moving available high yields cultivars and production technologies in different states over past few decades. There is a common observation transfer of technology from lab to land is very weak and many of these have not reached the ultimate growers. This may be one of the reasons for poor average sugarcane yield and sugar recovery as compared to both potentiality of sugarcane yield and recovery. Sugarcane and sugar output can be increased if the growers adopt the recommended package of practices relating to sugarcane production technology.

For increasing the level of adoption, farmers need to be convinced about recent knowledge regarding production technologies of sugarcane. In this regard, it is imperative to examine their status of knowledge and the factors which can influence of their adoption of new technology.

Useful study has done on various aspects of adoption of sugarcane production technology in different parts of country. However, very meager work has been done for adoption of sugarcane production technology among the tribal community of surguja division of Chhattisgarh. In view of the above mentioned facts, the present study entitled. **"A Study on Tribal Farmers of Surguja**

Division with Reference to Adoption of Sugarcane Production Technology" was carried out during 2013-14 with the following specific objectives:

OBJECTIVES

1. To study the socio - economic attributes of tribal sugarcane growers,
2. To ascertain the level of knowledge of the tribal sugarcane growers about recommended sugarcane production technology,
3. To find out the extent of adoption of sugarcane cultivation technology among selected tribal sugarcane growers,
4. To determine the relationship between characteristics of tribal sugarcane growers with extent of adoption regarding sugarcane production technology,
5. To study the constraints perceived by the tribal sugarcane growers and obtain their suggestions to overcome them.

The present study was conducted in Surguja division of Chhattisgarh state as it is most important sugar production division. Out of 13 blocks of both Surguja & Surajpur districts, 4 blocks (Batauli, Lundra, Surajpur and Pratappur) were selected purposively, because the maximum number of farmers in these blocks was involved in sugarcane production. From each block two villages (Total 2x4 = 8) were selected randomly and from each villages 16 tribal sugarcane growers (Total 8 x 16 = 128) have been selected randomly. In this way a total of 128 tribal sugarcane growers were selected as respondents for collection of data. The data were collected with the help of pretested interview schedule through personal interview. The major findings of the study are summarized under following sub-headings.

5.1 Independent variables

5.1.1 Socio- economic attributes

The data related to socio-economic condition have been presented in table 1, revealed that maximum number respondents (66.42%) were found to be in middle age group (35 to 55 years). The maximum respondents were under middle school (25.78%) while 21.09 per cent respondents under primary school, 21.87 per cent under illiterate, 17.18 per cent under higher secondary school, 10.15 per cent under high school and only (3.93%) of the respondents had college and above education level. The maximum number of the respondents (45.31%) had small size of land holding (1 to 2 ha.) followed by 23.43 per cent medium (2 to 4 ha). Out of the total occupied land, maximum 43.01 per cent sugarcane cultivated area occupied by the medium category of respondents. The majority of tribal sugarcane growers (40.62%) have more than 10 year of farming experience followed by 32.03 per cent with medium farming experience. The majority of respondents (83.59 %) belong to medium annual income (Rs. 27,232 to 1,99,324), followed by 14.07 per cent of respondents under high annual income and only 2.34 per cent of respondents under low annual income (Up to Rs. 27,231). Regarding social participation, maximum number of respondents (74.21%) had no membership in any organization followed by 16.45 per cent of respondents who were having membership in one organization. Only 5.47 per cent respondents who belonged to executive/ office bearer category while only 3.90 per cent had membership in more than one organization. The majority of the respondents (67.98%) had medium level of Scientific–orientation, followed by 18.75 per cent had high level of scientific–orientation.

5.1.2 Communicational characteristics of the respondents

About (58.59%) respondent had medium level of contact with extension agency contact, followed by 23.45 per cent respondents had low level of extension agency contact and only 17.96 per cent respondents had high level of extension agency contact. Results revealed that maximum number of the respondents *i.e.* 96.88 per cent contacted RAEOs weekly for getting new information related to sugarcane production technology. Respondents never contacted ADOs, SMS (KVK), Sugarcane Development officer (Sugar Factory), Agriculture Scientist (Agriculture University) and NGOs. Maximum number of respondents (71.09%) utilized medium level of information sources, progressive farmers (88.28 %), Radio (83.59%) and television (82.03 %) were important sources of information to be used for making the contact by the respondents for acquiring knowledge about recommended sugarcane production technology.

5.1.3 Knowledge of sugarcane production technology

The result revealed that majority of the respondent had low level of knowledge regarding selected 16 practices of sugarcane production technology *i.e.* disease management (100.00%), insect management (85.93%), tying of sugarcane (85.16%), seed treatment (75.78%), improved variety (29.69). Whereas, the majority of the respondents were having medium level of knowledge regarding sugarcane production technology *i.e.* fertilizer use (77.34%), improved variety (62.43%), weed management (41.41%), selection of land (39.06%), insect management (14.06), earthing up (12.05%), harvesting time (10.93%), seed rate (5.46). While in high knowledge level group selected practices is appropriate marketing facility (100.00%), preparation of land (98.46%), seed selection (96.88%), time of irrigation (95.32%), seed rate (94.53%), rationing management (93.75%), harvesting (89.06%), earthing up (87.05%), selection of land (58.59%), weed management (57.03%), seed treatment (21.88%), improved variety and fertilizer use (21.87%), tying of sugarcane (14.06%), of respondent were having high level of knowledge respectively.

5.2 Dependent variables

5.2.1 Adoption of sugarcane production technology

The maximum number of respondent (60.94%) had medium level of adoption of recommended sugarcane production technology i.e regarding field selection (71.87%), field preparation (98.43%), seed selection (98.43%), seed treatment (2.34 %), improved recommended varieties (53.90%), fertilizer application (87.05%) and insect pest management (12.37%). Whereas, in high level of adoption category the majority of respondent comes under regarding recommended seed rate (95.31%), time of irrigation management (87.5%), weed management (53.12%), earthing up (87.5%), Regarding harvesting time (64.84%), The majority of respondent (100.00%) indicate higher percentage of marketing, ratoon management(89.06%), respectively.

5.3 Correlation Analysis

It may be concluded from the result that out of all selected profiles characteristics, correlation coefficient between experiences, level of knowledge and Credit acquisition found to be positive and highly significant at 0.01 level of probability. However, the variable annual income and contact with extension agencies were found positive and significant relation with adoption at 0.05 per cent level of significance. The variables, age, occupation and cosmopolitans were found to be negatively correlation with adoption, while education, family size, land holding, scientific orientation and source of information showed no significant relation with extent of adoption of recommended sugarcane production technology by tribal sugarcane growers.

5.4 Multiple regression analysis

The result of multiple regression analysis revealed that out of 13 independent variables, level of knowledge contributed highly significant and Credit acquisition at 0.01 percent level of probability whereas experience and contact with extension agency were found significant to the adoption at 0.05 percent level of significance as evident from significant 't' value of these variables viz. experience, contact with extension agency and level of knowledge would cause 0.203, 0.155 and 0.477 unit changes, respectively in adoption of recommended sugarcane production technology by the tribal sugarcane growers. It could be further seen that all the 12 independent variables jointly explained the variation to the extent 46.3 percent in adoption of recommended sugarcane production technology, which was found statistically significant at 0.01 percent and 0.05 percent level of significant as could be seen form calculated 'F' value of 8.20.

5.5 Constraints faced by the sugarcane growers in adoption of recommended sugarcane production technology

Various problems which indicate that major constraints encountered by them under socio-economic constraints, damage of sugarcane crop by the wild animal (82.03%), lack of agriculture labour (71.09%), lack of risk bearing ability (57.03%), small size of land (45.31%), In case of social psychological constraints, the respondents reported to make not ready to adopt new technology (54.68%), Non-availability of information sources in proper time (50.78%), In case of technical constraints, maximum number of the respondents (92.18%) reported that lack of knowledge about insects identification and treatment, followed by lack of knowledge about of sugarcane tying (85.93%), non availability of electricity at proper time (71.87%), lack of knowledge about seed treatment methods (69.53%).

5.6 Suggestions of sugarcane growers for solving the constraints faced by them during the adoption of recommended sugarcane production technology

It is suggested that if the rate of adoption of recommended sugarcane production technology by tribal sugarcane growers wanted to be hiked there need to accelerate their frequency of contact with extension agency and level of knowledge through proper guidance and training. Other suggestion in order rank were, provide improved varieties by the government agencies, the crop lone and subsidy should be provided in proper time (83.59%), rate of seed, fertilizer and should be less (81.25%), electricity should be made available (78.12%) and provide subsidy fencing of farm or field by the government (69.53%) and demonstration should be conducted on farmers field by department of agriculture

(67.87%). This will be helpful for increasing the adoption of recommended sugarcane production technology.

Suggestions for action

The findings of the study reveals that majority of the sugarcane growers had medium level of knowledge regarding recommended sugarcane production technology. It was also observed that most of the respondents had medium level adoption of recommended sugarcane production technology. This indicates that the sugarcane growers were not aware about recommended sugarcane production technology and at the same time they were slow to adopt recommended sugarcane production technology.

The findings of the study showed that majority of the sugarcane growers came under medium level of adoption category; they were not using recommended technology fully. It is necessary to convince the sugarcane growers with the help of various extension teaching method like kisan mela, exhibition, group discussion, film shows and organization of demonstrations on improved technologies of sugarcane crop in the villages with the help of extension agents, uses of information sources in time and contact with other developmental organizations for providing the facilities and proper guidance to the sugarcane growers for adoption of recommended sugarcane production technology.

Conclusions:

The findings of the study indicated that majority of the tribal sugarcane growers belonged to middle age group, most of them had education up to middle school level, maximum number of them belonged to small size of family and majority of them had no membership in any organization. A significant majority of the respondents have been found to have no membership in any organization indicating very poor social participation.

Maximum number of respondents belonged to small size of land holding category, majority of them were involved in agriculture work as well as labour work for their occupation and majority of them had taken loan from nationalized bank and majority of respondents belonged to (up to 1 lakh) annual income.

Majority of the respondents came under the medium level of scientific–orientation category, majority of the respondents utilized medium level of information sources. Majority of the respondents were used neighbour/radio/progressive farmer/relative/friends frequently for seeking information about sugarcane production technology.

The findings of the study indicated that most of the tribal sugarcane growers were in middle level categories in respect to their extent of adoption regarding recommended sugarcane production technology. Thus, there is an urgent need to increase the extent of adoption of sugarcane growers about recommended sugarcane production, through proper utilization of source of information, extension contact, kisan mela and training programmes in different aspects should be conducted by the concerned agencies.

Majority of the sugarcane growers were having medium level of knowledge regarding recommended sugarcane production technology. Hence, extension efforts should be done to increase the level of knowledge of sugarcane growers about recommended sugarcane production technology.

From the result of correlation and multiple regression analysis it may be concluded that if the level of knowledge of sugarcane growers regarding recommended sugarcane production technology is to be expanded the extent of adoption will also be increased.

There is an urgent need to improve their education and knowledge level through providing education and improved sugarcane cultivation training, demonstration, field trips etc. The demonstration on use of various practices in sugarcane crop may therefore, help in persuade and changing the outlook of the sugarcane growers and exalting the extent of adoption of recommended sugarcane production technology.

Suggestions for future research work

On the basis of the results obtained from the study and the experience gained on completion of the investigation it is suggested that

1. As the numbers of independent variables were limited in the present research work a future study may be planned with more and different independent variables to know their contribution in adoption of recommended sugarcane production technology.

2. The study was limited to only 8 villages, four blocks and 2 district of Surguja division of Chhattisgarh state. Hence, a detailed study covering

more blocks and districts may be conducted in order to generalize the recommendations for the entire state of Chhattisgarh.

3. The role of sources of information in adoption of tribal sugarcane growers may be investigated in detail in order to make reliable suggestions for entire state.

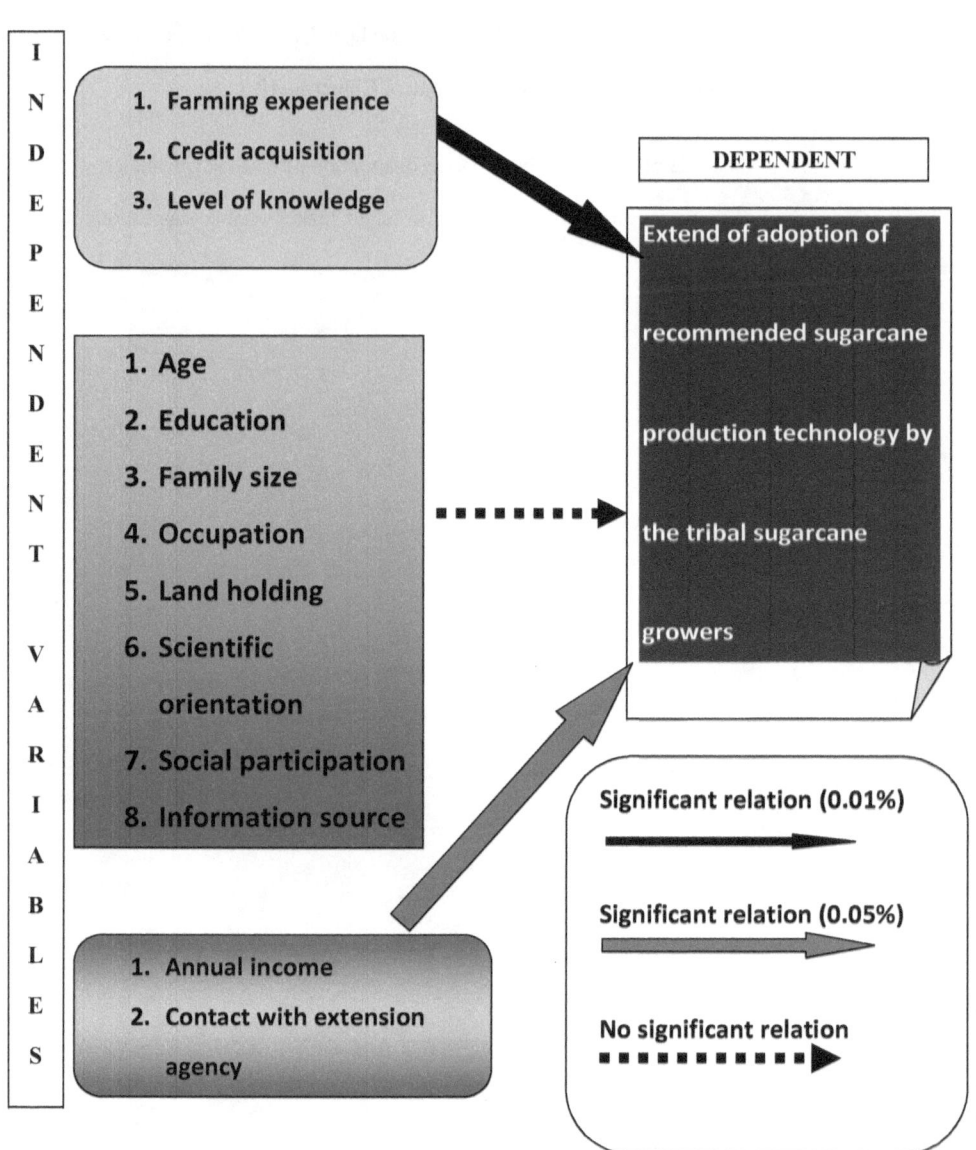

Fig. 5.1: Empirical Model of the study

Abstract

"A STUDY ON TRIBAL FARMERS OF SURGUJA DIVISION WITH REFERENCE TO ADOPTION OF SUGARCANE PRODUCTION TECHNOLOGY"

By

Rahul Kumar Tiwari

ABSTRACT

The present study was carried out at eight villages of Surguja & Surajpur district of Surguja division of Chhattisgarh state. Total 128 farmers were selected from the list of sugarcane growers as respondents and the data were collected through pretested interview scheduled and analyzed the data by using appropriate statistical tools & techniques.

The finding of present study revealed that the socio- economic profile of respondents belonged to middle age group, middle school level of education, small size of family composition, small size of land holding, highly occupied sugarcane land by the medium farmers, high level of farming experience, major source of income was agriculture with casual labour, maximum number of respondents belonged under low income group and early acquired credit for short term period from nationalized bank. Majority, of respondents had no member of any organization, medium level of scientific orientation, medium level of contact with extension agency and also had medium level of exposure to various sources of information for receiving the information about sugarcane cultivation practices from friends, relatives and neighbours etc.

The overall extent of knowledge and adoption of recommended production technology among respondents were found medium level (67.96 & 60.94%). Correlation coefficient between the selected attributes of the respondents with extent of adoption of recommended sugarcane production technology were observed as level of farming experience, knowledge of sugarcane production technology, annual income and contact with extension agency were found positive & significant. Remaining attributes of respondents i.e,

As for as major constraints in partial/ non adoption of recommended sugarcane production technology among respondents had identified as damage of sugarcane crop by the wild animal, insufficient agriculture labours, lack of motivation towards new technology, ignorant about insects identification and treatment.

The above problems should be solved with proper efforts of the extension machinery towards the tribal farmers to motivate them to adopt modern production technology of sugarcane in the study area.

Place: Raipur

Dr. P.K. Jaiswal
(Chairman Advisory Committee)

Bibliography

BIBLIOGRAPHY

Agarwal, G.C. and Goswami, H.G. 1992. Impact of cane cooperatives on sugarcane and adoption pattern of sugarcane growers. *Maharashtra Journal of Extension Education* **16**: 208-213.

Anonymous, 2013. Annual report (2011-12), Department of Agriculture, Government of chhatisgarh. http://agridept.cg.govt.in.

Aski, S.G., Dolli, S.S. and Sundaraswam, B. 1997. Impact of training on knowledge and adoption pattern of sugarcane growers. *Maharashtra Journal of Extension Education* **16**: 208-213.

Aski, S.K. 1989. Comparative study on knowledge and adoption behaviour of trained and untrained farmers. Farmers Training Centre (FTC), Arbhavi, *M.Sc. (Ag.) Thesis*, University of Agricultural Sciences, Dharwad (India).

Belligeri, S. B. 1996. A study on knowledge, adoption and perception of usefulness of agro-forestry practices by farmers of Hangal taluk, Dharwad district. *M. Sc. (Ag.) Thesis,* University of Agricultural Sciences, Dharwad (India).

Bhatkar, P. S., Shindhe, P.S. and Bhople, R. S. 1995. Influence of socio-economic and psychological factors on gain in knowledge by sugarcane growers. *Maharashtra Journal of Extension Education* **14**: 207-209.

Bhatkar, S.V. Shinde, P.S. and Bhople, R.S. 1996. Production technology constraints encountered by sugarcane growers in Vidarbha. *Maharashtra Journal of Extension* **15**: 86-89.

Bhatkar, S.V., Shinde, P.S. and Bhople, R.S. 1997. A study on correlates of adoption of sugarcane production technology by the farmers. *Maharashtra Journal of Extension Education* **16**: 18-21.

Bloom, S.D. 1979. Taxonomy of educational objectives: the classification of educational goals. *Handbook Incogetive Domin, Longman Group Ltd.* London (India).

Channal, G.P. 1995. A study on knowledge and adoption behaviour of share holders and non share holders of co-operative sugar factories in Belgaum district of Karnataka. *M.Sc.(Ag.) Thesis*, University of Agricultural Sciences, Dharwad (India).

Chauhan, S., Singh, S. R. K., Pande, A. K., and Gautam, U.S. 2013. Adoption dynamics of improved sugarcane cultivation in Madhya Pradesh. *Indian Research Journal of Extension Education* **13(2):** 26-30

Choudhary, N.M. 2003. Adoption behaviour of rural women regarding scientific storage practices of food grains in Raipur district of Chhattisgarh state. *M. Sc. (Ag.) Thesis*, IGKV, Raipur.

Dhruw, K. S. 2008. A study on adoption of recommended maize production technology among the farmers of Kanker district of Chhattisgarh state. *M. Sc. (Ag.) Thesis,* IGKV, Raipur.

Deshmukh, P. R., Kadam, R. P. and Sindhe, V. N. 2007. Knowledge and adoption of agricultural technologies in Marathwada. *Indian Research Journal of Extension Education* 7(1): 40-42.

Dubey, S.K., Sawanrkar, V.K.and Chakravarty, H.G. 1992. Knowledge and adoption of the rice production technology among small and marginal farmers. *Maharastra Journal of Extension Education* 11:79-84.

Goud, J.V. 1998 Sugarcane cultivation progress and prospects, Karnataka Institute of Applied Agricultural Research, Sameerwadi, Karnataka. P-4.

Gupta, S.P., Amardeep and Vir, K. 2003. Utilization of information sources by the farmers under different production system in Uttaranchal and Uttar Pradesh. *Manage Extension Research Review* 4(2):70-82

Itawdiya, K.K. and Singh, D.K. 2011. A study on technological gap in sugarcane production of Sehore block of Sehore district of Madhya Pradesh *M.Sc. (Ag.) Thesis*, JNKVV.

Kanavi, V.P. 2000. A study on the knowledge and adoption behaviour of sugarcane growers in Belgaum district of Karnataka. *M. Sc. (Ag.)* JNKVV, Jabalpur.

Karthikeyan, Chandrakandan, C. K. and Parvathi, S. 1996. A study of factors influencing Adoption behaviour of sugarcane growers in Pondichery Cooperative Sugar Mills Ltd., (PCSM) Pondicherry. *Maharashtra Journal of Extension Education* 15: 1331-1334.

Karthikeyan, C.; Subramaniyam, V.S. and Venkataprabhu 1995. Impact of registration on the socio-economic conditions of sugarcane growers. *Journal of Extension Education* **6**: 1049-1051.

Kathiresan, G.; R. Selvaraj; S. Ashokan; M.L. Manoharan; K. Duraisamy and M.R. Narayanswamy (2003). Adoption level and reason for non-adoption of improved technology of sugarcane by different categories of farmers. *Cooperative Sugar* **35**(2): 125-127.

Kharde, P.B. and Nimbalkar, S.D. 1986. Socio-economic factors affecting adoption of improved practices of sugarcane cultivation. *Maharashtra Journal of Extension Education* **14**.

Krishnamurthy, B., Mahadevaiah, D. Lakshminarayan, M.J. and Manjunath, B.N. 1998. Extent of adoption of recommended practices of sugarcane cultivation by farmers. *Journal of Extension Education* **9**: 2033-2036.

Khan, M.S, Krishna, T. and Rao, P.P 2002. Adoption pattern of eco-friendly technologies by rice growers. *Agricultural Extension Review* **3**(4): 22-25.

Kumar, A. and Singh, R. 2009. Wheat production constraints in Jharkhand. *Agricultural Extension Review,* January-March, pp. 26-30.

Lanjewar, O.P. 2009. Attitude of farmers regarding adoption of recommended cabbage production technology, with reference to use of drip irrigation system, in Durg and Raipur district of Chhattisgarh. *M. Sc. (Ag.) Thesis,* IGKV Raipur.

Limje, S. 2000. A study on adoption of recommended Soybean production technology among the farmers of Rajnandgaon district of M.P. *M. Sc. (Ag) thesis*, IGAU, Raipur (M.P.).

Mazher, Abbas, Sher, Mohammad, Nabi, Ifthikhar and Shiekh, A.D.2003. Farmer extension interaction and dissemination of recommended sugarcane production technologies in central Punjab (Pakistan). *International Journal of Agriculture and Biology* **5**(2): 134-137

Maraddi, G. N. 2006. A analysis of sustainable cultivation practices followed by sugarcane growers in Karnataka. *Ph.D Thesis,* University of Agricultural Sciences, Dharwad (India).

Mandal, P.K. 2000. Migration in the tribal area of Raipur city. M.A. Geography dissertation, Pt. Ravishankar Shukla University, Raipur (C.G.)

Mishra, A. 2006. A study on adoption of recommended Sugarcane Production Technology among the farmers of Kawardha district of Chhattisgarh. *M. Sc. (Ag.) Thesis,* IGKV, Raipur (C.G.).

Mukim, G.K. 2004. A study on adoption of recommended sunflower production technology among the farmers of Rajandgaon district of Chhattisgarh. *M. Sc. (Ag.) Thesis,* IGKV, Raipur.

Nagaraj, M. V. 2002. A study on knowledge of improved cultivation practices of sugarcane and their extent of adoption by farmers in Bhadra Command area in Davangere district, Karnatka. *Ph.D Thesis,* University of Agricultural Sciences, Dharwad (India).

Naik, R.D. 2005. A study on knowledge and adoption pattern on improved sugarcane practices in Bidar district of Karnataka state, *M.Sc.(Ag.) Thesis,* University of Agricultural Sciences, Dharwad (India).

Natikar, K. V., 2001, Attitude and use of farm journal by the subscriber farmer and their profile. A critical analysis. *Ph. D. Thesis,* University of Agricultural Sciences, Dharwad (India).

Palaniswamy, A. and Sriram, N. 2001. Modernization characteristics of sugarcane growers. *Journal of Extension Education* **11**(4): 2906-2915.

Pal, S.K., Halim, M.A., Kashem, M.A. and Karim, A.Z. 2001. Factor influencing the adoption of recommended cultural practices in sugarcane farming in Ishurdi areas of Bangladesh. *Chiang Mai Journal of Science* **9** (1): 61-82.

Pandey, P.K. Suryawanshi, D.K. and Sarkar, J.D. 2004. Credit acquisition pattern of rice grower in C.G. *In IRRI (abstract)* p. 270.

Prajapati, I.B. 2006. A study on socio-economic factors responsible for technological gap of recommended wheat technology among tribal farmers of Sidhi district (M.P.). *M. Sc. (Ag.) Thesis,* JNKVV, Jabalpur.

Patel, M.M. Chatarjee, A. and Sharma, H.M. 1994. Knowledge and adoption level of sugarcane growers. *Maharashtra Journal of Extension Education* **13**: 131-14.

Patel, M.K. 2008. A study on technological gap in recommended soybean production technology among the farmers of Kabirdham district of Chhattisgarh state. *M. Sc. (Ag.) Thesis,* IGKV, Raipur.

Pawar, P.P. Jadale U.D. Shinde, C.B. and Kale, P.V. 2005. Adoption of improved production technology on sugarcane farms in Western Maharashtra. *Cooperative Sugar* **36**(10): 805-812.

Raghavendra, M.R., 2004. Knowledge and adoption level of post harvest technologies by redgram cultivators in Gulbarga district of Karnataka. *M. Sc. (Ag.) Thesis*, Univ. Agric. Sci., Dharwad, India

Raghuvanshi, H.S. 2005. Adoption behaviour of rice growers regarding control measures of various insect pests of rice crop in Dhamtari district of Chhattisgarh state. *M. Sc. (Ag.) Thesis,* IGKV, Raipur.

Rajni, T. 2006. Impact of mushroom production and processing training on farm women organized at Indira Gandhi Agricultural University, Raipur (C.G.). *M. Sc. (Ag.) Thesis*, IGKV, Raipur.

Ramsingh and Singh, S.B. 1994. Input management in sugarcane cultivation. *An Advances in Agricultural Research In India* **1**: 28-29.

Sharma, T.N., Singh, R.K. and Sharma, V.B. 2000. Effect of adoption on the adoption behavior of trained farmers of Krishi Vigyan Kendra, Chhindwara, Madhy Pardesh. *JNKVV Research Journal* **34** (1&2)

Sharma, R.R. 2001. A comparative study of adoption of the improved selected package of practices for rice and wheat crops in Gohad block of Bhind district (M.P.). *M. Sc. (Ag.) Thesis*, JNKVV, Jabalpur.

Shivanand, P. 2007. Human resource development activities initiated by Nandi sugar factory Bijapur district of Karnataka state, *M.Sc. (Agri) Thesis*, Uni. of Agril. Sci., Dharwad (India).

Singh, K. V. Kumar S. and Malik, R.N. 1993. Resource use efficiency in sugarcane farmers (A micro level study of Western Uttar Pradesh). *Cooperative Sugar* **25**: 43-45.

Singh, R. and Kumar, A. 2007. Weed control strategies adopted by farmers in wheat crop. *Agricultural Extension Review*, July-Dec. pp. 13-15.

Singh, D., Singh, B.K. and Singh, R.L. 2007. Constraints in adoption of recommended rice cultivation practices. *Indian Research Journal of Extension Education* **7**(1): 70-73.

Shrivastava, K.K., Trivedi, M.S. and Lakhera, M.L. 2002. Knowledge and adoption behaviour of Chilli growers. *Agricultural Extension Review* **7**(8): 22-25.

Suryawanshi, R.K. 2009. A study on adoption of finger millet production technology by the tribal farmers of Bastar district of C.G. *M. Sc. (Ag.) Thesis,* IGKV, Raipur.

Talior, R.S., Pande, A.K. and Sanoria, Y.C. 1998. Socio-Personal Correlates of knowledge and adoption of farming practices of farmers of watershed area. *Maharastra Journal of Extension Education* **1** (1): 20-25

Tomar, R.S. 1993. A study of the technological gap in adoption of package of practices of wheat cultivation among the farmers of Porsa block district Morena (M.P.). *M. Sc. (Ag.) Thesis,* JNKVV, Jabalpur.

Vekaria, R.S., Pandya, D.N. and Padgeria, M.M. (1990). Role of cooperative sugar factories in sugar cane development. *Maharashtra Journal of Extension Education* **9**: 301.

Verma, S. 2009. A study on knowledge and adoption of organic farming practices in paddy cultivation among the tribal farmers of Kanker district (C.G.). *M.Sc. (Ag.) Thesis,* IGKV, Raipur.

Appendix

"छत्तीसगढ़ राज्य के सरगुजा संभाग में आदिवासी किसानों द्वारा गन्ने की अनुशंसित उत्पादन तकनीक के अंगीकरण के अध्ययन हेतु साक्षात्कार अनुसूची"

साक्षात्कार अनुसूची

प्रश्नावली क्र. ....

परामशदार्ता	शोधकर्ता का नाम
डॉ. पी. के. जायसवाल	राहुल कुमार तिवारी
प्राध्यापक	एम. एस. सी. कृषि अंतिम वर्ष
कृषि विस्तार विभाग, इं.गां.कृ.वि.वि.	कृषिविस्तार विभाग इं.गां.कृ.वि.वि
रायपुर (छ.ग.)	रायपुर (छ.ग)

1. जिला   :   ....
2. विकासखण्ड   :   ....
3. ग्राम   :   ....
4. कृषक का नाम   :   ....
5. ग्राम की दूरी   :   मुख्य मार्ग से .... शहर से ....
6. कृषक की उम्र .... 35 वर्ष तक ☐   35–55 वर्ष ☐   55 वर्ष से अधिक ☐
7. शिक्षा का स्तर
    1. अशिक्षित   ☐
    2. प्रायमरी   ☐
    3. मिडिल स्कूल   ☐
    4. हाईस्कूल   ☐
    5. हायर सेकेन्ड्री   ☐
    6. स्नातक व अधिक   ☐
8. आपके परिवार का आकार
    1. छोटा परिवार 1–5 सदस्य   ☐
    2. मध्यम परिवार 6–10 सदस्य   ☐
    3. संयुक्त परिवार 10 से अधिक सदस्य ☐

9. कृपया आप अपने ग्राम में कार्यरत संस्थाओं एवं उसमें अपनी सहभागिता के बारे में निम्न जानकारी दीजिए :

क्र. सं.	संस्थायें	भागीदारी (हॉ/नही)	सदस्य	पदाधिकारी
1.	ग्राम पंचायत			
2.	सहकारी समिति			
3.	युवा मंडल			
4.	सांस्कृतिक मंच			
5.	स्कूलजनभागिदारी समिति			
6.	महिला मंडल			
7.	किसान क्लब			
8.	अन्य 1. 2. 3.			

11. आपका मुख्य व्यवसाय क्या है?

 1. कृषि
 2. कृषि+ मजदूरी
 3. कृषि+ नौकरी
 4. कृषि+ पशुपालन
 5. कृषि+ व्यापार
 6. कृषि+ अन्य ....

12. आपकी विभिन्न श्रोतों से कुल वार्षिक आय कितनी है ?

क्र. सं.	व्यवसाय	वार्षिक आय (रूपये में)
1.	कृषि	
2.	पशुपालन	
3.	मजदूर	
4.	व्यापार	
5.	नौकरी	
6.	अन्य 1. 2. 3.	
	कुल वार्षिक आय (रूपये में)	

13 आपके पास कृषि योग्य कुल कितनी भूमि है

भूमि .... (एकड़ / हे. में)

    1. सीमांत कृषक (1.0 से 2.0) एकड़   ☐

    2. लघु कृषक (2.0 से 4.0) एकड़   ☐

    3. मध्यम कृषक (4.0 से 10.0) एकड़   ☐

    4. बड़ा कृषक (10 से अधिक) एकड़   ☐

14 (अ) सिंचाई के साधन

    1. नहर   ☐
    2. नलकूप   ☐
    3. कुआ   ☐
    4. तालाब   ☐
    5. अन्य   ☐

(ब) सिंचाई की उपलब्धता

    1. निरंक   ☐
    2. आंशिक (25% तक)   ☐
    3. मध्यम (25.1–50% तक)   ☐
    4. अधिक (50.1–75% तक)   ☐
    5. बहुत अधिक (75.1% से अधिक)   ☐

15 औसत सिंचित भूमि

    1. खरीफ में   ....

    2. खरीफ और रबी में   ....

    3. बहुवर्षयी में   ....

16. आप कौन – कौन सी फसलों का उत्पादन किया है।

क्र.	मृदा प्रकार	फसल	फसल की किस्म	बुआई की विधी	सिंचित		असिंचित	
					क्षेत्र (ह. मी)	उत्पादन	क्षेत्र (ह. मी)	उत्पादन
1.		खरीफ फसल 1................. 2................. 3.................						
2.		रबी फसल 1................. 2................. 3.................						
3.		जायद फसल 1................. 2................. 3.................						

17. गन्ना फसल कब से उगा रहे है ? ..सन्...................

18 क्या आपने ऋण लिया है ? (हॉं / नही) यदि हॉं तो बताइये किन स्त्रोतों से ऋण प्राप्त किया है –

क्रं.	स्त्रोत	ऋण प्राप्त किया (हां / नही)	ऋण अवधि	ब्याज दर	ऋण उपलब्धता	
					सरलता से	कठिनाई से
1.	सहकारी संस्था					
2.	राष्ट्रीयकृत बैंक					
3.	साहूकार					
4.	मित्र					
5.	पड़ोसी					
6.	रिश्तेदार					
7.	अन्य					

19 आप अपने आसपास के शहर अथवा ब्लॉक से कितना संपर्क करते है?

1. कभी नहीं
2. बहुत कम ;माह में एक बार
3. कभी-कभी ;सप्ताह में दो बार
4. रोज/नियमि

20 निम्न लिखित कथनों से आप कितना <u>सहमत/असहमत</u> है ? कृपया अपने विचार बतलाइये।

क्र. सं.	कथन	सहमति का स्तर		
		पूर्ण	आंशिक	नरंक
1.	एक किसान को आर्थिक लाभ एवं पैदावार के लिये खेती करना चाहिए।			
2.	जो किसान सबसे अधिक लाभ कमाता है वही सबसे अधिक सफल माना जाता है ।			
3.	किसान को सिर्फ नई तकनीक अपनीनी चाहिए जो सबसे अधिक लाभ दें ।			
4.	किसानों को ज्यादा लाभ कमाने के लिए अनाज वाली फसलों के बजाय नगद फसल लेनी चाहिए ।			
5.	किसी भी नये काम में किसान के बच्चों की अच्छी शुरूआत करना कठिन होता है जब तक उन्हे वह आर्थिक सहायता न दें ।			
6.	सभी को जीवन में कमाई करना चाहिये मगर पैसा कमाना सब कुछ है ऐसा नही मानना चाहिये ।			

21 आप नई तकनीकी अथवा पद्धति को अपनाने से संबंधित निम्न लिखित कथनो से किस स्तर तक सहमत है?

सं.	विचार	पूर्णत सहमत	सहमत	असहमत	कुछ नहीं कह सकते	अपूर्ण सहमत
1.	नई पद्धति को जल्दी अपनाने से कोई हानि नहीं होती।					
2.	किसान को नई पद्धति तब तक नहीं अपनाना चाहिए जब तक कि उसकी क्षमता के बारे में जानकारी न हो।					
3.	नई पद्धति को अपनाने से किसानों में समृद्धता आती है।					
4.	नई पद्धतियों को अपनाने से फसलों की उत्पादकता मे वृद्धि होती है।					
5.	नई पद्धतियों किसानों की आवश्यकताओं को पूरा करती है एवं समस्याओं का समाधान होता है।					
6.	नई पद्धतियों को तब तक नही अपनाना चाहिए जब तक किसान स्वयं उनकी क्षमता न देख ले।					
7.	नई पद्धतियों को अपनाना सदैव जोखिम एवं अनिश्चितता से भरा होता है।					

22 आपको आप गन्ना उत्पादन की तकनीकी से संबंधित जानकारी किन स्त्रोतों से प्राप्त होती है –

कं.	स्त्रोत	जानकारी			विश्वस्नीयता का स्तर		
		अक्सर	कभी–कभी	कभी नहीं	पूर्ण	आंशिक	निरंक
1	उन्नत कृषक						
2	पड़ोसी						
3	मित्र						
4	रिश्तेदार						
5	ग्रामीण नेता						
6	सहकारी समिति						
7	कृषि पत्र–पत्रिकाएं						
8	रेडियो						
9	टी.वी.						
10	किसान मेला						
11	प्रशिक्षण						
12	किसान भ्रमण						
13	लीफलेट / पाम्पलेट						
	अन्य 1. .... 2.						

23 क्या आप गन्ना उत्पादन की तकनीकी से संबंधित जानकारी के लिये कृषिविस्तार कार्यकर्ताओं से संपर्क करते है । (हॉ/नही) यदि हॉ तो कृपया निम्न जानकारी दीजिए—

क्र. सं.	कृषिविस्तार कार्यकर्ता	संपर्क का अंतराल (वर्ष में)			
		कभी नही	वर्ष मे (1–3 बार)	माह मे एक बार	सप्ताहिक
1.	ग्रामीण कृषिविस्तार अधिकारी				
2.	कृषिविकास अधिकारी				
3.	वरिष्ट कृषिवैज्ञानिक				
4.	विषय वस्तु विशेषज्ञ				
5.	कृषिवैज्ञानिक				

24 निम्न कथनों के आधार पर गन्ने की फसल के उत्पादन के बारे में जानकारी देवें ।

क्र.	कथन	सहमति का स्तर		
		पूर्ण	आंशिक	निरंक
1.	क्या आपको गन्ने की फसल के लिये उपयुक्त भूमि के चुनाव संबंधी जानकारी है । ;हां/ नहीं यदि हां तो विवरण दे 1 2 3			
2.	क्या आप गन्ने की फसल के लिये भूमि की तैयारी से अवगत है ;हां/ नहीं यदि हां तो विवरण दें 1 2 3			

3.	क्या आपको गन्ने की फसल के बीज चयन संबंधी जानकारी है । ;हां/ नहीं यदि हां तो विवरण दें 1 2 3			
4.	आप गन्ने की फसल में बीजोपचार के लिये किन–किन दवाओं का उपयोग करते है । क्या आपको इसकी जानकारी है ;हां/ नहीं दवाओं के नाम मात्रा 1 2 3			
5.	आप गन्ने की बुआई किस पद्धति से करते है एवं कितनी बीज दर का उपयोग करते है । बुआई पद्धति बीज दर ;कि.ग्रा./ए. 1 2 3			
6.	क्या आपको गन्ने की फसल के विभिन्न प्रकार से उपयोग हेतु उन्नत किस्मों की जानकारी है । हां/ नही यदि हां तो विवरण दे अ. गुड़ के लिये उन्नत किस्में 1 2 3			

	ब. शक्कर के लिये उन्नत किस्मे 1 2 3			
7	क्या आपको मालूम है कि गन्ने की फसल में खाद तथा उर्वरक की कितनी मात्रा का उपयोग करते है। हां / नहीं यदि हां तो विवरण दें खाद गोबर / जैविक मात्रा कि.ग्रा. / एकड़ 1 2 3 उर्वरक मात्रा कि.ग्रा. / ए 1 2 3			
8.	क्या आपको गन्ने की फसल के लिये आवश्यक एवं उपयुक्त सिंचाई के समय की जानकारी है । हां / नहीं यदि हां तो विवरण दे सिंचाई का समय ऋतु अंतराल दिन 1. 2. 3.			

9.	क्या आपको गन्ने की फसल में खरपतवार नियंतरण की जानकारी है । हां / नहीं यदि हां तो विवरण दे 1. 2. 3.			
10	क्या आपको गन्ने की फसल में कीटनाशकों के छिड़काव संबंधी जानकारी है । हां / नहीं यदि हां तो विवरण दें कीटनाशी का नाम मात्रा कि. ग्राम 1. 2. 3.			
11.	क्या आपको गन्ने की फसल में लगने वाले रोगों तथा उनके नियंत्रण विधि की जानकारी है। हां / नहीं यदि हां तो विवरण दें रोग दवाई मात्रा कि.ग्रा / ए. 1. 2. 3			
12.	क्या आपको गन्ने की फसल में मिटटी चढ़ाने की विधि से अवगत है । हां / नहीं यदि हां तो विवरण दें 1. 2. 3.			

13.	क्या आपको गन्ने की फसल में बंधाई की विधि के बारे में जानते हैं । हां / नहीं यदि हां तो विवरण दें 1. 2. 3.			
14.	क्या आपको गन्ने की फसल पकने एवं कटाई के उपयुक्त समय की जानकारी है । हां / नहीं यदि हां तो विवरण दें 1. 2. 3.			
15.	क्या आपको गन्ने की फसल के विपणन संबंधी जानकारी है। हां / नहीं यदि हां तो विवरण दें 1. 2. 3.			
16	क्या आप को गन्ने की पेडी प्रबंध की जानकारी है 1 2 3			

25 गन्ने की फसल के लिये अनुशंसित तकनीक का अंगीकरण

क्र.	कथन	अंगीकरण का स्तर		
		पूर्ण	आंशिक	निरंक
1.	आप गन्ना किस प्रकार की भूमि में लगाते है । भूमि का प्रकार 1. 2. 3.			
2.	आप गन्ने के लिये भूमि की तैयारी कैसे करते है । विधि 1. 2. 3.			
3.	आप गन्ने के लिये किस प्रकार के बीजों का चयन करते है । बीज का प्रकार 1. 2. 3.			
4.	आप गन्ने के लिये बीजोपचार किस प्रकार करते है । फफूंद नाशक मात्रा 1. 2. 3			

5.	आप गन्ने की बुआई किस पद्धति से करते है एवं कितनी बीज दर का उपयोग करते है । बुआई पद्धति बीज दर कि.ग्रा/ए. 1. 2. 3.			
6	आप गन्ने के विभिन्न प्रकार से उपयोग हेतु आने वाली कौन– कौन सी उन्नत किस्मों को लगाते है । गुड़ के लिये किस्म उपज कि.ग्रा/ए. 1. 2. 3. शक्कर के लिये किस्म उपज कि.ग्रा/ए. 1. 2. 3. रस के लिये किस्म उपज कि.ग्रा/ए. 1. 2. 3.			

7	आप गन्ने की फसल में खद एवं उर्वरक की कितनी मात्रा डालते है। खाद गोबर/वर्मीकम्पोस्ट मात्रा;कि.ग्रा/ए. 1. 2. 3. उर्वरक मात्रा कि.ग्रा/ए. 1. 2. 3.			
8.	आप गन्ने की फसल में कब सिंचाई करते है । सिंचाई का समय 1. 2. 3.			
9.	आप गन्ने में खरपतवार नियंत्रण के लिये कौसी विधि अपनाते है । विधि मात्रा कि.ग्रा/ए. 1.निंदाई–गुढ़ाई 2.मशीनों द्वारा 3.शाकनाशी द्वारा			
10.	आप गन्ने में कीट नियंत्रण किस प्रकार करते हैं । विधि/कीटनाशक मात्रा 1. 2. 3.			

11.	आप गन्ने में रोग नियंत्रण किस प्रकार करते हैं । विधि/रोगनाशक मात्रा 1. 2. 3.			
12.	आप गन्ने में मिट्टी कब एवं किस प्रकार करते है। विधि मात्रा 1. 2. 3.			
13.	आप गन्ने की बंधाई कब एवं कैसे करते है । विधि 1. 2. 3.			
15	आप गन्ने की फसल का किस तरह एवं कहां विपणन करते है । 1. 2. 3.			
16.	आप किस प्रकार गन्ने की पेडी प्रबंध करते है 1 2 3			

26. आपको गन्ने की अनुशंसति तकनीक को अपनाने में निम्नलिखित में से किन कठिनाईयों का सामना करना पड़ता है। (हॉ/नही)

क्रं.	कठिनाईयाँ	
1.	**व्यक्तिगत समास्या** 1. कम पढ़ा लिखा होना 2. उम्र में अधिकता 3. परिवार बड़ा होना 4. दूसरा व्यासाय होना 5. भूमि जोतों का बंटा होना 6. अन्य.....	☐ ☐ ☐ ☐ ☐
2.	**समाजिक –आर्थिक समास्या** 1. भूमि कम होना 2. आय का कम होना 3. भूमि बंटी होना 4. ऋण समय पर न मिलना 5. ऋण पर्याप्त मात्रा में प्राप्त न होना 6. उच्च ब्याज दर 7. जोखिम सहन करने की क्षमता का न होना 8. सामाजिक प्रोत्साहन में कमी 9. अन्य लोग उन्नत कृषि तकनीक नहीं अपनाते 10. कृषि श्रमिकों की कमी 11. जानवरों द्वारा फसलों को हानि .... 12. कृषि कार्य में मजदूरी दर का ज्यादा होना 13. अन्य...	☐ ☐ ☐ ☐ ☐ ☐ ☐ ☐

3.	सामाजिक – मनोवैज्ञानिक समस्या	☐
	1. सामाजिक भागीदारी कम होना	☐
	2. नई तकनीक अपनाने के लिये तुरंत तैयार करना	☐
	3. रुढ़ी वादिता	☐
	4. अन्य	☐
	5.	
4.	सूचना संबधी समस्या	
	1. सूचना समय पर न मिलना	
	2. सूचना पर्याप्त मात्रा में न मिलना	
	3. सूचना माध्यम का उपलब्ध न होना	
	4. कृषि वस्तिार अधिकारी व वैज्ञानिको का समय पर उपलब्ध न होना	
5.	तकनीकी करण	
	1. कृषि साधनो बीज उर्वरक कृषि यंत्र कीटनाशक का समय पर उपलब्ध न होना ।	
	2. उन्नत तकनीकी के विषय में पूर्ण ज्ञान का अभाव	
	3. उन्नत तकनीकी के बीजोपचार विधि का मालूम न होना	
	4. बिजली उपलब्धता में कमी	
	5. गन्ने की फसल में दिये जाने वाले खाद एंव उर्वरक की मात्रा के विषय में ज्ञान का अभाव	
	6. गन्ने की बंधाई के विषय में ज्ञान की कमी	
	7. गन्ने की फसल में लगने वाले कीटों की पहचान एंव प्रबंधन में पूर्ण ज्ञान न होना	

6.	परिवहन संबधी समस्या 1. .... 2.	
7.	विपणन संबधी समस्या 1. .... 2.	

25. अनुशंसित तकनीकों को अपनाने में आये व्यवधानों को दूर करने के लिए उपाय बताइये ।

....
....
....
....
....
....
....
....
....
....
....
....
....

www.ingramcontent.com/pod-product-compliance
Lightning Source LLC
Chambersburg PA
CBHW020915180526
45163CB00007B/2741